# ON THE QUANTA EXPLANATION OF VISION

# ON THE QUANTA EXPLANATION OF VISION

## PROEFSCHRIFT

TER VERKRIJGING VAN DE GRAAD VAN DOCTOR IN DE WIS- EN NATUURKUNDE AAN DE RIJKS-UNIVERSITEIT TE UTRECHT, OP GEZAG VAN DE RECTOR MAGNIFICUS DR. H. WAGENVOORT, HOOGLERAAR IN DE FACULTEIT DER LETTEREN EN WIJSBEGEERTE, VOLGENS BESLUIT VAN DE SENAAT DER UNIVERSITEIT IN HET OPENBAAR TE VERDEDIGEN OP MAANDAG 7 MAART 1949, DES NAMIDDAGS TE 4 UUR

DOOR

## MAARTEN ANNE BOUMAN

GEBOREN TE UTRECHT

Springer-Science+Business Media, B.V.

1949

ISBN 978-94-017-5718-8    ISBN 978-94-017-6063-8 (eBook)
DOI 10.1007/978-94-017-6063-8

PROMOTOR: PROF. DR J. M. W. MILATZ

AAN MIJN OUDERS
AAN TREES

# VOORWOORD

Bij de voltooiïng van dit proefschrift betuig ik gaarne mijn dank aan allen, die tot mijn wetenschappelijke vorming hebben bijgedragen.

Allereerst, U Hooggeleerde Milatz, Hooggeachte Promotor, ben ik dankbaar voor de wijze, waarop U mij in staat gesteld hebt de onderwerpen, die mijn belangstelling hadden te bewerken. Zeer in het bijzonder ben ik U erkentelijk voor Uw daadwerkelijke steun mij geboden in moeilijke omstandigheden voortvloeiend uit de oorlogs-toestand.

U, Hooggeleerde Weve, dank ik voor de belangstelling, die U steeds voor mijn werk hebt getoond. Dat ik een gedeelte van de experimenten voor dit proefschrift als medewerker van de Organisatie voor Zuiver Wetenschappelijk Onderzoek verrichtte, was mede aan Uw medewerking te danken.

Zeergeleerde Fischer, de gedachtenwisselingen, die ik met U heb mogen voeren, hebben veel bijgedragen voor het verkrijgen van dieper inzicht in zintuig-physiologische problemen. Uw hulp voor het verschijnen van dit manuscript heb ik op hoge prijs gesteld.

Beste van der Velden en ten Doesschate. De onderlinge discussies, wederzijdse hulp en samenwerking waren voor mij zeer waardevol.

Tenslotte dank ik alle medewerkers van het Physisch Laboratorium voor de vriendschap en medewerking.

# ON THE QUANTA EXPLANATION OF VISION

BY

## M. A. BOUMAN

11-V-'48

**Survey of contents.**

# INTRODUCTION.

Though much knowledge has been accumulated on the sense of vision it is only recently [1, 2, 3]) that without formulating any hypothesis concerning the inner mechanism of visual excitation a quantitative explanation of the problems of liminal excitation was put forward.

In the past, several authors [4, 5]) have checked the possibility of explaining the behaviour of several visual functions as visual acuity, contrast-sensitivity, a.s.o. by the mass action law. The general idea of these photochemical theories is that the visual impression is conditioned by the concentration of a substance produced by the decomposition of the photochemical material of the rods and cones. In this photochemical theory several parameters arbitrarily chosen are used so as to make the theory fit the experiments. Most of these parameters are connected with the properties of photochemical and chemical reactions of the substances of the rods and cones. The values of these parameters are not fundamental constants of the visual processes as it proved to be necessary to give them different values for the explanation of experiments on the same eye, even when these experiments refer to related visual functions as for instance the several functions for rod vision. By this reason the photochemical theory is far from being satisfactory. It is a pity that the real existence of the chemical systems chosen ad hoc by the theories hitherto proposed are not satisfactorily established by chemical means.

Since the photochemical explanations especially those advanced by Hecht [4]) are widely known many investigators [5]) have tried to check and extend these theories, whilst the results of simple experiments were not satisfactorily interpreted by the photochemical theory. Indeed, the behaviour of absolute threshold values as a function of the time of observation and as a function of the visual angle of the testspot, nor the flickering effect visible at the threshold of vision in an illuminated field can be explained by this theory. In 1932 Bowling Barnes and Czerny [6]) developed that the flickering effect mentioned is due to the statistical fluctuations of the light quanta. So, they discovered the influence of the nature of the stimulating energy on the behaviour of the visual perception. This influence is a pure physical aspect of vision.

The existence of the various aspects of vision is widely recognized but some authors have over-emphasized to considerations lying in their particular field of science. A physicist is inclined to pure physical considerations like a psychologist to psychological, a.s.o.

No doubt, this onesidedness is also partly the result of our scanty knowledge of the fundamental facts concerning the chemical, physiological and histological constitution of the eye, so that the theory-creator is given free play. These facts are demonstrated by the existence of photochemical explanations neglecting influences of nerve-physiological, psychological and physical kind.

From a psychological point of view, the important issue is the process in the brain leading to conscious luminous perception. By psychological reason a quantitative treatment of the behaviour of light-impressions is possible only by means of just distinguishable differences between impressions, whether of colour, brightness or hue, the quantitative information obtained by the ascertaining of equality included. Indeed, we can only give a qualitative indication of the difference between impressions.

When subjective measurements are carried out by means of successive or simultaneous comparison of visual impressions, the difficulty-arising from the poor power of memory for impressions or the poor ability for judging impressions lying far apart in the field of view-is a psychological aspect of vision. This difficulty is met with by giving suitable values to the time or space interval in the visual field of the test person. From this it is clear that, in general, the influence of changes of time or visual angle on the results of our experiments can be of psychological origin.

The structure of the retina, optical nerve and brain will also give us indications concerning the way a luminous perception arises. Polyak[7]) and Østerberg[8]) have contributed largely to our knowledge of the histology of the retina and of the connections of the rods and cones with the optic nerve fibers. The constitution of the optic nerve system is however such that for the time being but little concerning function can be predicted from it. Only this is certain, that it plays the part of intermediary between the excitation of the retina and the conscious luminous perception. It is, therefore, to be expected that our growing knowledge of the histology and properties of this system, will help to solve the problems of the behaviour of the visual functions. In chapter IV we will discuss some of these properties, whereas here as an example Hartline's[9]) electro-physiological investigation may be mentioned. He found that a fiber of the optic nerve is stimulated through the excitation of the receptors of a well-defined area of the retina; it therefore follows, reversely, that the separate receptors within such an area can mutually influence each other via their connections with the fiber mentioned of the optic nerve. Such influence will certainly express itself in the behaviour of the resolving power or visual acuity and other visual functions.

Apart form the structure, the physiological mechanisms by which the nerve system propagates the stimuli, is sure to play an important part. Though we know as yet but little of this, a few facts are available, which can be regarded as sufficiently reliable to serve as a foundation for a theory concerning the way a luminous perception arises.

Foremost among these is the "all or none law". When a fiber reacts on an excitation impulses are transmitted by the fiber; the size and shape of the separate impulses are independent of the kind and intensity of the excitation. So, the smallest reaction of a fiber is the transmission of one impulse. The corresponding intensity of excitation is the threshold value for reaction. For lower intensities no impulses are transmitted at all. For higher intensities the number of impulses increases for increasing intensity.

When we try to form an idea about the way a sensation arises we must not break this all or none law. For instance, this law predicts that when a fiber reacts on different kinds of excitation of the sense organ, the reaction of this separate fiber can only give information of the intensity and not of the kind of excitation. So, the intelligence of the separate fiber is rather poor.

Of course, the properties of the photochemical substances and the chemical reactions in the rods and cones are important too. Especially, Wald [10]) and Bliss [11]) has contributed largely to our knowledge in this field. It is clear that the sensitivity of the eye as a function of the wave-length of the incident light must be related to the corresponding function of the absorption power of the substances mentioned. The properties of the chemical reactions will influence the behaviour of the adaptation and other visual functions.

The treatment of the influence of the physical aspect of vision by Bowling Barnes and Czerny [6]) was only qualitative. In 1943, de Vries [12]) explained the behaviour of contrast sensitivity as a function of intensity with statistical considerations on the number of quanta absorbed in the rod and cones. In 1944, the fluctuating results in the observation of light flashes at the threshold of rod vision were studied quantitatively by van der Velden. [1]) He explained by a statistical treatment the fluctuation phenomena and the general behaviour of the threshold values by means of the two quanta theory without involving any arbitrary parameter and he found some fundamental properties of the nerve system. In these experiments on light flashes of various visual angles and times of observation it appeared to be possible to distinguish between the physical and physiological influence. It can be expected that by a consistent extension of such experiments the several aspects of vision can partly be isolated.

The experiments described in this thesis were performed to check the two quanta theory by repeating the experiments on rod vision of van der Velden. Besides this we consistently extended these experiments for the analysis of many other problems.

In chapter I the two quanta theory, its fundamental experiments and our experimental confirmation are discussed, the work of investigators of experiments of the same kind being considered. Chapter II gives an exposition of threshold experiments for rod vision in which a very large number of combinations of visual angle and flash time are studied. Chapter III deals with the investigation on cone vision. In chapter IV some general properties of the nerve system are discussed. In chapter V experiments on the spectral sensitivity curve for rod and cone vision are presented, whereas in chapter VI a study of the brightness impression is given. In chapter VII the behaviour of visual acuity is explained with the aid of the newly acquired knowledge. The last chapter is devoted to the general aspects of vision.

# I. THE TWO-QUANTA EXPLANATION.

## 1. The experiments of van der Velden and the theoretical foundation of the two-quanta explanation.

For the determination of the smallest amount of energy necessary for vision it is obvious to measure the energy of a test spot that is recognized when it is watched for some time. However, it is known that the larger the area observed, the larger need the threshold energy be for its recognition as far as the size exceeds a certain value. Below this size, according to Ricco's law, the threshold energy is independent of the size. By this reason the visual angle of the test spot must be chosen within this region of independence.

For the dependence on the time of observation the behaviour is similar. Bunsen's law predicts that the energy at the threshold is independent of the time of observation as far as the time is small. Outside this region this energy increases with increasing time. So, it is clear that the time too have to be small. Resuming: we must use short flashes with a small angular size.

In the observation of these flashes fluctuating results are found. A flash originating from a constant light source is sometimes seen and sometimes not. For the determination of the threshold energy, van der Velden [1]) studied this phenomenon and it proved that the region of intensities in which the fluctuations occur is rather large. Therefore he passed over to the measuring of *the chance of observation* for several intensities of the flashes. (See figure 1).

In the region mentioned the chance changes from 0 tot 100 % and he defined the threshold by the energy corresponding to a fixed chance of observation, namely 60 %.

In figure 2 we present his experimental arrangement.

A tungsten ribbon filament lamp a was used as a light source. In order to get a sufficiently low brightness two opal glasses c and e were placed into the beam. The size of the light spot was controlled by interchangeable diaphragms f. A Schott Vg2 filter b with maximum transmission at 5300Å determined the colour of the beam.

The head of the observer was fixed by means of a mouthpiece mounting. The place of the retina where the flash was received was about 7° nasal from the fovea of the dark-adapted right eye. This fixation was realized with the aid of a red fixation light observed by the left eye. The aperture of the pupil of the right eye was reduced to a constant size by the intercalation of a diaphragm h.

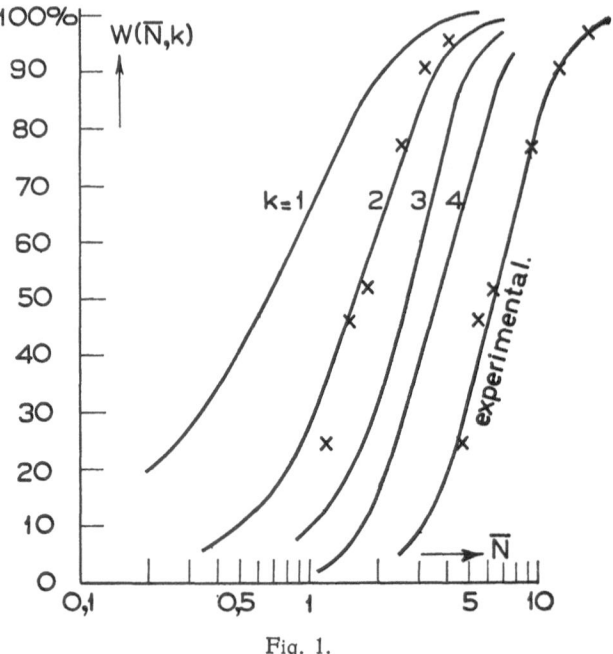

Fig. 1.

The chance of observation as a function of the average number of quanta of a flash $\overline{N}$ for the cases $k = 1, 2, 3\ 4$, when $f = 1$. The experimental results for one of the test persons are presented. (flash-time $< 0{,}01$ sec, visual angle $= 4'$).

The duration of the flash could be varied by means of a compur shutter $g$, which was regularly checked with the aid of an alternating light source, photo-cell, amplifier, and oscillograph. The properties of the opal glasses were studied with a photo-cell. The transmission

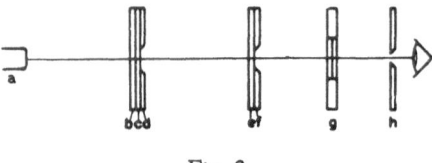

Fig. 2.

curve of the filters had been determined with the photoelectric a.c. amplifier with a.c. galvanometer described by Milatz [13]) and Milatz and Bloembergen. [14]) The tungsten ribbon filament lamp is calibrated according to the principles developed and used by Ornstein and co-workers [15]) in our institute.

16

The number of quanta per second incident on the eye for every wave-length is further provided by the strength of the current of the lamp and the geometry of the experimental arrangement. The intensity of the flash was varied by changing the distance and the current of the lamp.

### A. Determination of the type of the relation between the chance of observation and the intensity of the flash.

In figure 1 the curve representing the experimental results on the chance of observation as a function of the intensity of the flashes was represented. Each measured chance is obtained with abouth 50 repetitions of a flash of fixed intensity.*)

It proved that the number of quanta incident on the eye necessary for a chance of observation of 60 % is rather few. The data range from 8 to 28 quanta of light.

When flashes originating from a constant light source are presented to an observer statistical fluctuations in the actual number of quanta of the flashes occur. Van der Velden considered the influence of this statistical behaviour of the number of quanta on the observation of light flashes and he explained quantitatively the fluctuating phenomenon in the observation of flashes with the aid of the statistical behaviour mentioned.

In the past, the fluctuating phenomenon at the threshold of vision was often thought to be due to the physiological or psychological variability of the sense organ. It now proved to be caused by a pure quantum-physical aspect.

The initial process of the visual light perception is the absorption of light quanta in the photochemical material of the receptors of the eye. Let us now make the assumption that light is only perceived when $k$ or more than $k$ quanta of the flash are absorbed in the visual purple. From the light incident on the eye is only a part absorbed by the visual purple in order to cause a light impression; let this fraction be $f$. All losses of light caused by reflections and absorption in the various parts of the eye are considered in this fraction. The chance of observation will depend on the intensity of the flash indicated by the average number of quanta $\overline{N}$ and on the number of quanta $k$, that must be at least absorbed in the visual purple in order to perceive light. Therefore we indicate this chance with $W(\overline{N},k)$.

---

*) Error in the measured chance $W \%$ for $p$ repititions is about $\dfrac{\sqrt{p \cdot W(1-W)}}{p}\,\%$.

With the aid of Poisson's formulae mathematical considerations learn that

$$W\,(\overline{N},\,k) \;=\; 1 \;-\; \sum_0^{k-1}{}^s \frac{(f\,\overline{N})}{s\,!} \; e^{-f\,\overline{N}} \tag{1}$$

Plotting $W(\overline{N},k)$ against $\log \overline{N}$, the shape of this curve is independent of the value of $f$, but different for the various cases $k = 1, 2, 3$, etc. (See figure 1).

Comparing the experimental curve with the theoretical curves $f$ and $k$ can be determined. It was found that in three test persons the experimental curves coincided best for the case $k = 2$. $f$ was 28, 9 and 7 %.

So the study of the fluctuating phenomenon at the threshold of vision leads to a method for the determination of the number $k$. We will indicate this method further on by method $A$.

Van der Velden developed two other independent approaches to the establishing of the number of quanta absorbed in the visual purple, necessary for the light perception. All these three methods are independent of the value of $f$; *So the losses of light in the various parts of the eye and the absolute values of the energies employed are of no importance. Even the wearing of tinted glasses will not influence the result concerning k!* Also the possibility that not all quanta reaching the retina will be absorbed so as to contribute to the light perception have no influence. Such an effect cannot be distinguished from the other losses of light in the eye and is also included in the fraction $f$.

The two other methods, described under $B$ and $C$, were based on a determination of the threshold energy as a function of the flashtime and visual angle of the test spot.

*The conclusion drawn from the experimental results was the two-quanta explanation: when two quanta are absorbed within an area corresponding to a visual angle D (about 10') and within a time τ (about 0,02 sec) a light impression results.*

As according to Østerberg [8]) the number of rods in an area of 10 minutes of arc in this part of the retina is about 100 the chance is negligeable that the two quanta are absorbed in one rod. *By this it is evident that a rod reacts on the absorption of one quantum and transmits an impulse to its nerve-connection. When two rods within an area of 10' and within about 0,02 seconds give such an impulse a light-impression is received.*

## B. Determination of the Dependence of the Average Number of Quanta for a Fixed Chance of Observation on the Duration of the Flash for a Small Visual Angle.

As the experiments concerning $W(\overline{N}, k)$ could reasonably well be described by the hypothesis that two quanta are necessary for a light perception, it is obvious that one has to expect a dependence of the average number of quanta in the flash $\overline{N}$, necessary for a certain chance of observation, on the duration of the flash $t$. The retinal condition existing after the absorption of the first quantum will have a finite lifetime $\tau$. If the second quantum is absorbed after this time there will be no light impression at all.

For the time of observation $t > \tau$ an extra condition was introduced in formula (1) for $k = 2$, expressing the necessity that during the flash at least two quanta follow each other within $\tau$.

When $W(\overline{N}, t)$ is the chance of observation of the flash of $t$ seconds it was found [1,2]

$$W(\overline{N}, t) = 1 - e^{-f\overline{N}} \sum_{s=0}^{m+1} \frac{(f\overline{N})^s}{s!} \left\{ 1 - \frac{(s-1)\tau}{t} \right\}^s ; \quad m\tau \leq t \leq (m+1)\tau \quad (2)$$

When $t \leq \tau$ this equation becomes the same as (1) and constitutes the theoretical foundation of Talbots law. When $t \gg \tau$ we can derive a formula that gives a fair approximation. The chance that the time between two absorptions is between $T$ and $T + dT$ is

$$\frac{f\overline{N}}{t} e^{-f\overline{N} \cdot T/t} dT.$$

The chance that light is perceived by the two absorptions is

$$\int_0^\tau \frac{f\overline{N}}{t} e^{-f\overline{N} \cdot T/t} dT = 1 - e^{-f\overline{N} \cdot \tau/t}$$

The chance for no light perception at all will be about

$$\left( e^{-f\overline{N} \cdot \tau/t} \right) f\overline{N} = e^{-(f\overline{N})^2 \cdot \tau/t},$$

as the average number of pairs of succeeding quanta is $f\overline{N} - 1$ and $f\overline{N} \gg 1$. So that

$$W(\overline{N}, t) = 1 - e^{-(f\overline{N})^2 \cdot \tau/t}. \quad (3)$$

For a chance of observation of 60 % $\overline{N}_{60}$ % was calculated as a function of $t/\tau$ for $f = 1$ with the aid of equation (2). Plotting $W(\overline{N}, t)$ against log $\overline{N}_{60}$ % the shape of this function is independent of the value of $f$.

Obviously $\overline{N}_{60}$ % is constant for the case $k = 1$.

When $k = 2$ it appears from equations (2) and (3) that $\overline{N}_{60}$ % is proportional to $t^{\frac{1}{2}}$ for $t \gg \tau$. For the case that three quanta must be absorbed in the visual purple within a time $\tau$, $\overline{N}_{60}$ % is proportional to $t^{\frac{2}{3}}$ for $t \gg \tau$, analoguous to $k = 3$ under C. See figure 3.

19

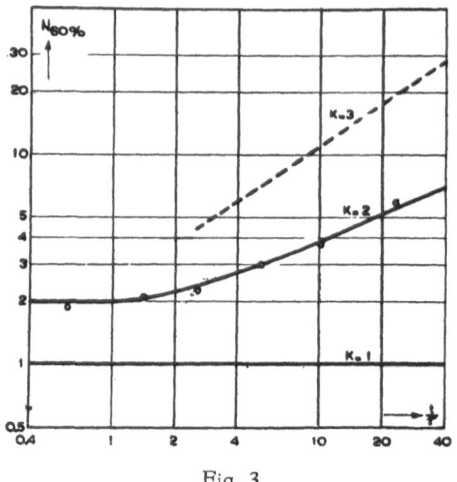

Fig. 3.

The theoretical number of quanta, necessary for a chance of observation of 60 % as a function of $t$, the duration of the flash, for the 1, 2, and 3 quanta hypothesis. The experimental resul's for $d < D$ from Fig. 8a reduced to $f = 1$ are presented.

Corresponding differences for probability curves representing form (1) are much smaller. The experimental results of van der Velden concerning $\overline{N}_{60\,\%}$ for various values of $t$ were satisfactorily rendered by the 2-quanta case. In view of the great differences in the slope with $t$ of $\overline{N}_{60\,\%}$ for the cases $k = 1, 2, 3$ is impossible to conclude to one or three quanta. $\tau$ proved to be about 0.02 sec. The visual angle of the test spot was about 4 minutes.

### C. Determination of the Dependence of the Average Number of Quanta for a Fixed Chance of Observation on the Visual Angle of the Flash for a Short Time of Observation.

In the case $k = 2$ it is obvious that one has also to expect a dependence of $\overline{N}_{60\,\%}$ on the visual angle. The condition caused by the absorption of a quantum will have an extension corresponding to the sensitive unit $D$ of the retina. If the second quantum is absorbed outside the sensitive unit activated by the first quantum no light is perceived. As long as the image of the light spot is small compared with the sensitive unit there will only be a slight dependence of $\overline{N}_{60\,\%}$ on the visual angle $d$ of the flash.

Let $W(\overline{N}, d)$ be the chance of observation of a flash with visual angle $d$. When $d \ll D$ it was found

$$W(\overline{N}, d) = 1 - e^{-f\overline{N}}(1 + f\overline{N}) - (f\overline{N})^2 e^{-f\overline{N}}\left(\frac{5}{6}\frac{d}{D} - \frac{3}{8}\frac{d^2}{D^2}\right) \tag{4}$$

When $d \rightarrow 0$ this equation merges into (1).

The chance for the flash falling completely within one sensitive unit and the chances for the falling of a definite fraction within one unit and a definite fraction within an adjoining unit are multiplied by the corresponding chances for a light perception. These results are added, and in this way (4) is obtained. It was assumed that the sensitive units as well as the light spot were square. For circular units the results will be only slightly different.

The course of $\overline{N}_{60}$ % with $d$ according to (4) is only of slight importance as the deficiencies of the optical system of the eye obliterate its influence when $d < 2$ minutes.

When $d \ll D$,

$$W(N, d) = 1 - \left[e^{-f\overline{N}.\,D^2/d^2}\left(1 + f\overline{N}.\frac{D^2}{d^2}\right)\right]^{d^2/D^2} \tag{5}$$

as $\quad e^{-f\overline{N}.\,D^2/d^2}\left(1 + f\overline{N}.\frac{D^2}{d^2}\right)$

is the chance for no light perception of a sensitive unit. The number of sensitive units is $d^2 D^2$. $N_{60}$ % was calculated as a function of $d$ by means of (4) and (5). Plotting $W(\overline{N}, d)$ against $\log \overline{N}$ the shape of this function is again independent of the value of $f$. As appears from (5), $\overline{N}_{60}$ % is proportional to $d$ for $k = 2$ and $d \gg D$. For (5) can be reduced when $d \gg D$,

$$W(\overline{N}, d) = 1 - \left[\left\{1 - f\overline{N}.\frac{D^2}{d^2} + \frac{(f\overline{N})^2}{2!}\frac{D^4}{d^4} - \ldots\right\}\left(1 + f\overline{N}.\frac{D^2}{d^2}\right)\right]^{d^2/D^2}$$

as $\quad f\overline{N}.\frac{D^2}{d^2} \overset{\prime}{<} 1 \quad$ so that

$$W(\overline{N}, d) = 1 - e^{-1/2\,(f\overline{N})^2.\,D^2/d^2} \tag{6}$$

It can also be proved that $\overline{N}_{60}$ % is proportional to $d^{2(k-1)/k}$ for arbitrary values of $k$ when $d \gg D$. Analogous $W(\overline{N}, t)$ is proportional to $t^{(k-1)/k}$ when $t \gg \tau$.

For the case $k = 1$, $\overline{N}_{60}$ % is independent of $d$, so that again great differences occur in the slope of $\overline{N}_{60}$ % with $d$ for $d \gg D$, for $k = 1, 2, 3$. The experimental results of van der Velden were satisfactorily rendered by $k = 2$. $D$ proved to be 10 minutes. The flash time was about 0.01 sec. (see Fig. 4).

Equations (4) and (5) constitute the theoretical foundation of the laws of Ricco and Piper.

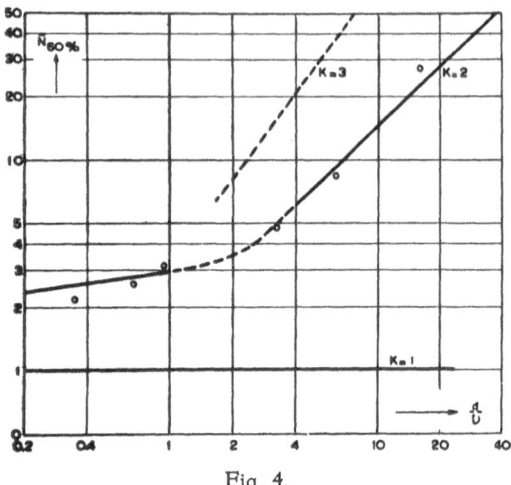

Fig. 4.

The theoretical number of quanta necessary for a chance of observation of 60 % as a function of the visual angle $d$ of the
light spot. The theoretical values are given for the cases, that 1, 2
and 3 quanta are necessary for the perception of light. The experimental results for $t < \tau$ from Fig. 7b reduced to $f = 1$ are given.

## 2. Theoretical and experimental observations on the two-quanta explanation.

Independently from us Hecht and his co-workers [16]) performed
an investigation partly of the same kind.

Only a fraction of the quanta incident on the eye is absorbed by
the visual purple so as to give an impression of light. About half
of the amount is lost by reflection and absorption in the various
parts of the eye. Moreover not all quanta reaching the retina will be
absorbed so as to contribute to the light perception. Hecht determined this effective fraction reaching the retina with the aid of
threshold values for various wave-lengths * and concluded that as
an upper limit 5—14 quanta must be absorbed in the visual purple.
His experimental results point to 1 quantum as the lower limit. From
this part of Hecht's work it can be concluded, that the number of
absorbed quanta necessary for the light perception is at least one
and does not exceed 14.

For the exact determination of this number Hecht used only the
method described under $A$ in Section 1, applying a visual angle of

---

* See page 65.

10 minutes and a time of about 0.004 seconds. From the shape of the resulting curve he concluded that the cooperation of at least 5, 6, or 7 quanta (dependent on the observer) were necessary for the perception of light which disagrees with the results of van der Velden and those described in this paper according to which independent of the observer two quanta are sufficient.

For several reasons the two-quanta explanation appears to be the better founded.

Hecht and his co-workers used the same eye for the measurement of threshold values and for the observation of the red light for the fixation of the eye. It is however obvious [1,2] that the red fixation light affects the observation of the light flashes, especially when the chance of observation is small. Consequently one obtains under these conditions a steeper probability curve than in the case that care is taken to observe the flash of light and the fixation light separately as was done in our experiments. Without this precaution the required number of quanta will obviously turn out to be higher.

For the exact determination of the number $k$ method $A$ described in Section 1 and used by Hecht is less suitable compared with the methods $B$ and $C$. Besides the smaller usefulness caused by the smaller differences in the slope of the curves for the various cases $k = 1, 2, 3, \ldots$ one can make objections of a theoretical nature against the method as soon as $k$ is proved to be not equal to one.

It was assumed for all formulae of Section 1 when two quanta are absorbed in the visual purple the chance for a light perception is independent of the time $\triangle T$ between the two absorptions when $\triangle T \leq \tau$ and is zero when $\triangle T > \tau$. It is very probable that this is only approximately correct and a more continuous decrease of this chance $W(\triangle T)$ for $\triangle T$ values near $\tau$ will occur for increasing $\triangle T$.

It is easy to understand that the slopes of the curves of method $A$ will be steeper for a definite value of $k$ by this matter. The chance for a light perception for a continuous course of $W(\triangle T)$ increases faster with increasing $\overline{N}$ of the flash as the chance for a light perception increases when the average time between the absorptions of the quanta decreases.

It was also assumed for the formulae of Section 1 that when two quanta are absorbed in the visual purple the chance for a light perception is independent of the distance between the two absorptions when the two quanta fall within a sensitive unit of diameter $D$. Besides this according to (4) the slope of the curve representing the chance function of method $A$ depends on the visual angle of the flashes and is always steeper than equation (1) predicts. It is very probable that the chance for a light perception is not independent of the distance $\triangle d$ between two absorptions of the two quanta fallen within a sensitive unit $D$, analogous to the facts for the dependence of the time between the two absorptions. By this fact the slope of the curve of method $A$ is again steeper. (See fig. 5).

So the number $k$ obtained from the experiments described in method $A$ with the aid of formula (1) is the upper limit of the real number $k$. *Only when the slope of the curve agrees with the case $k = 1$ it is evident that these difficulties do not exist and no further information is needed [17]).*

It is not possible to conclude to the actual existence of definitely bounded sensitive units in the retina from the dependence of the slope of the chance function on the visual angle according to equation (4). It might be possible that the two quanta have to be absorbed within a distance $D$ of each other instead of the necessity of two absorptions within an area of diameter $D$ in the retina.

Only the continuous decrease of the chance for a light perception with increasing distance of the absorptions of the two quanta $W(\triangle d)$ can cause the dependence of the slope of the chance functions of Fig. 5 on the visual angle. The formulae (4), (5), and (6) are derivated for the case that the sensitive

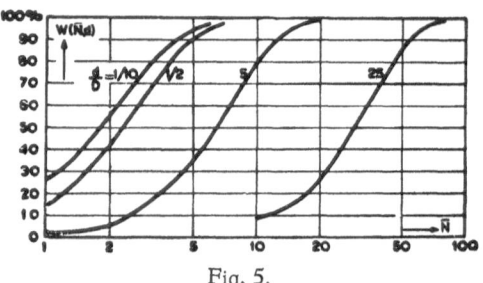

Fig. 5.

Chance of observation as a function of $\overline{N}$ according to the two-quanta hypothesis for different values of $d/D$ and $f = 1$ from the formulae 4 and 5.

units did exist in the retina. When for the light perception two quanta must be absorbed within a distance $D$ of each other a formula similar to (2) is obtained. As the problem is now two-dimensional $d^2$ corresponds to $t$, and the slope of the threshold values $\overline{N}_{60\%}$ with $d$ when $d \gg D$ is not influenced by this matter. For that reason we cannot conclude to 'he existence of sensitive units from the experiments of the methods $A$, $B$, or $C$.

*The number $k$ obtained from the curve of method $A$ will be closest the real number $k$ when the visual angle of the light spot and the flash time are as small as possible. The influence of the facts just mentioned is in these conditions the smallest.*

Hecht [16]) applied a visual angle of 10 minutes and a flash time of about 0.004 seconds. As this visual angle is not small compared with $D$, the number of quanta necessary for the light perception was bound to come out too high. The influence of the visual angle on the slope of the curve of method $A$ is also demonstrated by the experiments of C. Peyrou and H. Piatier, [18] who concluded to the value 2—5 for $k$. The deviations in their data from the theoretical two-quanta slope of equation (1) are still greater than ours, even for the smallest visual angle of 2 minutes. The deviations in our data almost agrees with equation (4). Unfortunately the authors did not give information about the manner of fixation of the eye and the spot of the retina on which the flashes were received. In Hecht's work the flash was seen 20 degrees nasal from the fovea, in ours 7 degrees.

It is very probable that the condition of the nerve system and the spot of the retina used in the observation of the flashes influence the chance for a light perception when two quanta are absorbed as function of the time and distance of these two absorptions: $W(\triangle T, \triangle d)$. These facts can cause a variability in the resulting curves of method $A$.

The two-quanta explanation was not only corroborated by the probability curves for the observation of light flashes of method $A$. Indeed, methods $B$ and $C$ are much more valuable for an exact determination of $k$, and for the test person of the present investigation, the results for $t < 3\tau$ and $d < 2D$ can only be explained by the two-quanta case, not only when method $A$ was applied but also when methods $B$ and $C$ were applied.

It can be shown that for a continuous course of the chance for a light perception $W(\triangle T, \triangle d)$ the slope of the threshold curves $\overline{N}_{60\,\%}$ of the methods $B$ and $C$ as far as $d \gg D$ and $t \gg \tau$ are still quite similar to the corresponding slopes of equations (2), (3), (5), and (6), which were derivated for the chance function $W(\triangle T, \triangle d)$ represented by the dotted line of Fig. 6.

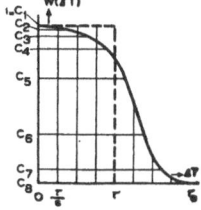

Fig. 6.

Chance for a light perception when two quanta are effectively absorbed as a function of the time $\triangle T$ between the absorptions.

We give the derivation for the time relation. For the visual-angle relation the facts are analogous. For an arbitrary course with time represented by the full line of Fig. 5, we divide the segment $\tau_0$ in a large number of equal parts $\tau_0/s$ with corresponding average value of $W(\Delta T, \Delta d)$: $c_k$. The chance that the time between two succeeding absorptions is between $\Delta T$ and $\Delta T + d(\Delta T)$ is

$$\frac{f\bar{N}}{t} e^{-f\bar{N}\cdot\Delta T/t} d(\Delta T)$$

The chance for a light perception by these two absorptions is

$$\int_0^{\tau_0/s} \frac{f\bar{N}}{t} e^{-f\bar{N}\cdot\Delta T/t} c_1 d(\Delta T) + \int_{\tau_0/s}^{2\tau_0/s} \frac{f\bar{N}}{t} e^{-f\bar{N}\cdot\Delta T/t} c_2 d(\Delta T) + \ldots$$

$$+ \int_{(s-1)\tau_0/s}^{\tau_0} \frac{f\bar{N}}{t} e^{-f\bar{N}\cdot\Delta T/t} c_s d(\Delta T) =$$

$$= 1 - (c_1 - c_2) e^{-\frac{1}{s}\frac{\tau_0}{t}f\bar{N}} - (c_2 - c_3) e^{-\frac{2}{s}\frac{\tau_0}{t}f\bar{N}} - \ldots - c_s e^{-\frac{\tau_0}{t}f\bar{N}}$$

As the average of succeeding pairs of quanta is $f\bar{N} - 1$, the chance for no light perception at all is about

$$\left[ (c_1 - c_2)(1 - \frac{1}{s}\frac{\tau_0}{t}f\bar{N}) + (c_2 - c_3)(1 - \frac{2}{s}\frac{\tau_0}{t}f\bar{N}) + \ldots + c_s(1 - \frac{\tau_0}{t}f\bar{N}) \right]^{f\bar{N}} =$$

$$= \left[ c_1 - \frac{1}{s}\sum_{k=1}^{s} c_k \frac{\tau_0}{t}f\bar{N} \right]^{f\bar{N}}$$

so that

$$W(\bar{N}, t) = 1 - c_1^{f\bar{N}} \cdot e^{-\frac{1}{s}\sum_{k=1}^{s}\frac{c_k}{c_1}\frac{\tau_0}{t}(f\bar{N})^2}$$

when $\tau_0 \ll f\bar{N}/t$, analogous to form 3 as $c_1$ can always be chosen equal to one therefore, *only the methods B and C are suitable for the exact determination of the number k.*

It is evident, that the presented derivation is not right for every course of $W(\Delta T, \Delta d)$ with time $\Delta T$ or distance $\Delta d$. One of the conditions is that these functions are non-increasing. As soon a maximum (optimal) value in these functions occur there must be a region of time-values or visual angles of the flashes within which the threshold-values increase with decreasing flash-time or visual angle. Indeed there would be a definite time-lag or distance between the two absorptions for which the chance for a light-perception is maximal. *)

However, a behaviour of the threshold-values of this kind is not found, either as a function with time, or as a function with the visual angle. Talbots law (it=constant) holds for all values of time

---

*) This is not probable in view of the nervous mechanisms described in chapter IV.

as far as $t<0,1$ sec. and Ricco's law ($id^2 = $ constant) holds for all values of $d$ as far as $d$ does not exceed 2'-15', dependent on the place of the retina used for the observations.

There is still one more reason for the probability that the conclusion of Hecht for the necessity of 5, 6, or 7 quanta is too high. For one of the observers [1, 2, 3] was found $f = 28$ percent at $\lambda = 5050$Å, on the suppositions that the two-quanta hypothesis is correct and that the course of the threshold values with wave-length is always in agreement with the international scotopic luminosity curve. Be that as it may, for $d = 4$ minutes, $t = 0.01$ seconds, and $\lambda = 5050$Å, for this observer $\overline{N}_{60 \, 0/0}$ never did exceed 14 (see page 35). Since about half the amount of light incident on the eye is lost by reflection and absorption in its various parts, the upper limit of the average number of quanta reaching the retina is 7. Only a part of these quanta will be absorbed in the visual purple so as to give a light perception. If 5, 6, or 7 quanta would be necessary this fraction would have to be improbably large, as Wald [19] found this fraction to be of the order of magnitude of 20 percent.

In our experiments the absolute value of the energy is of no importance for the determination of the number $k$. Only the value of $f$ of the test person is based on the absolute values. As we determined these absolute level in an indirect way, the error may be large but will certainly not exceed 25 percent. For most of the test persons $f$ has under optimal conditions a value of 7 to 10 percent, so that $\overline{N}_{60 \, 0/0}$ when $t<\tau$ and $d<D$ is about 25 quanta. As far as they proved to be reliable the corresponding numbers in the work of other investigators are for the greater part almost twice as large. Hecht [16] examined all the published measurements on the energy required for vision. The data of Chariton and Lea yield 18 quanta, von Kries and Eyster give 34—68, Barnes and Czerny's work results in 40-87, Hecht found 58-148, and Peyrou and Piatier yield 70-90 quanta. The differences may be partly due to the variability of the absorbing and reflecting properties of the human eye. Moreover, the experimental arrangements of the several authors were not the same. In this connection the location of the illuminated area of the retina, the way of fixating the eyes and the general comfort of the observer are of special importance. In Pirenne's [20] continuation of the work of Hecht, experiments were performed in order to find out if there is any cooperation between the two eyes for the perception of light. His experiments for the left and

right eye are not sufficiently complete to determine the number of quanta necessary for the perception of light. Besides this, Pirenne has used white light, so that his measurements are not strictly comparable to ours. He found that there was no cooperation between the two eys. This is in agreement with our result that the two-quanta have to be absorbed within a distance $D(10')$ of each other in one eye.

De Vries' estimation [12]) of the number of quanta to be absorbed at the threshold of vision is not very convincing, since the results of different authors are used. The relation derived by him between contrast sensitivity and brightness for low values of the intensity using the one-quantum hypothesis, is not in accordance with the experiments of Steinhardt. [21])

Steinhardt found for low brigthness values proportionality between the contrast sensitivity and the brigthness, not the square root of the brightness, whereas de Vries derived proportionality with the square root of the brightness.

The experimental results described in the next chapter for the subject M.A.B. also agree completely with the two-quanta hypothesis as far as the flas-time $t < 3\tau$ or the visual angle $d < 2D$.

Accepting the results of the two-quanta theory Baumgardt's investigation [22]) represents the experimental confirmation given by us as the base for the two-quanta conclusion. His experiments according the methods $B$ and $C$ for the determination of the number $k$ agree with ours and are an independent confirmation of the two-quanta theory.

## II. THE DEVIATIONS FROM THE TWO-QUANTA EXPLANATION FOR THRESHOLD-VALUES WHEN BOTH THE FLASH-TIME AND THE VISUAL ANGLE OF THE LIGHT-SPOT ARE LARGE.

### 1. Our confirmation of the two-quanta theory when t<τ or d<D.

In chapter I it was shown that the behaviour of threshold-values as a function of the flash-time when the visual angle is small and as a function of the visual angle when the flash-time is small can be explained completely by the two-quanta theory.

In this chapter we will present some further experiments relating to the behaviour of the threshold values when the flash time $t > \tau$ together with the visual angle $d > D$.

The experimental arrangement was the same as described in the previous chapter.

For many combinations of values of $t$ and $d$ the chance for observation, $W(\overline{N}, t, d)$, of the flash was determined as a function of $\overline{N}$. The values of $\overline{N}$ for an observation chance of 60 % was determined from these measurements and is again called $\overline{N}_{60\,\%}$. Using the formulae (2) and (5) we obtain the theoretical expression for $W(N, t, d)$ for arbitrary values of $t$ and for $d >> D$ *)

$$W(\overline{N}, t, d) = 1 - \left[ e^{-f\overline{N}, D^2/d^2} \cdot \sum_{s=0}^{m+1} \frac{(f\overline{N}.D^2/d^2)^s}{s!} \left(1 - (s-1)\tau/t\right)^s \right]^{d^2/D^2} \quad (7)$$

in which $m\,\tau \leq t \leq (m+1)\,\tau$.

The theoretical values of $W(\overline{N}, t, d)$ and $\overline{N}_{60\,\%}$ were derived from equations (7) and (4), which are also good for small values of $d$.

The theoretical and experimental values of $W(\overline{N}, t, d,)$ and $\overline{N}$ are given in Figs. 7a-d and the results for $\overline{N}_{60\,\%}$ in Figs. 8a and 8b.

*On repeating the experiments of van der Velden [1, 2] with another test person, we found again that the experimental $W(\overline{N})$ curve, in the case that t<τ and d<D, was in accordance with the two-quanta explanation.*

---

*) The chance for no light perception of a sensitive unit for small times mentioned in the derivation on page 21 is in equation (5) replaced by this chance for arbitrary times.

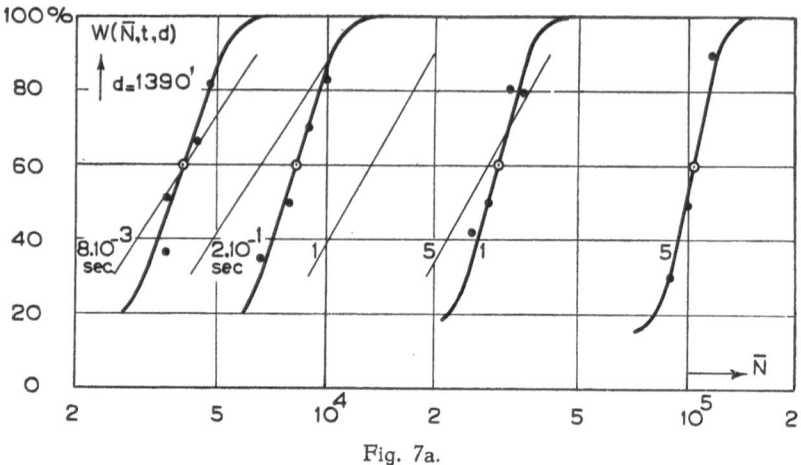

Fig. 7a.

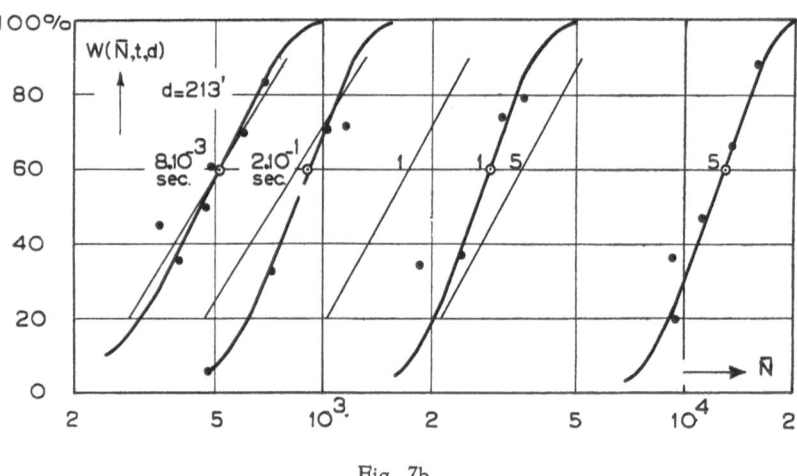

Fig. 7b.

Both the thresholds as a function of $d$, when $t < \tau$ and the thresholds as a function of $t$ when $d < D$ are also fully covered by the two-quanta case, as can be seen from Figs. 8a and 8b. The slopes of the curves mentioned permit to conclude to $k = 2$ with an accuracy of about 0,1. For the test person (M.A.B.) used in these experiments we found $D = 12'$, $\tau = 0.035$ sec. and $f = 8$ percent. The value of $\tau$ is a little higher than for H. v.d. V. in the previous experiments ($\tau$ about 0.02 sec.).

30

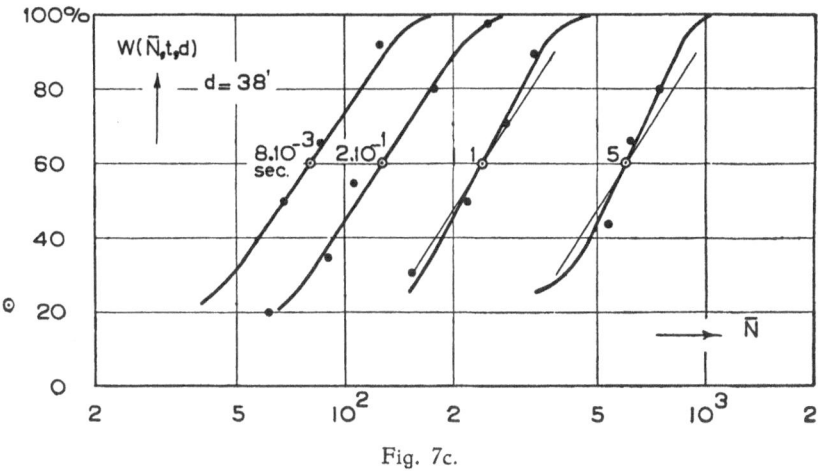

Fig. 7c.

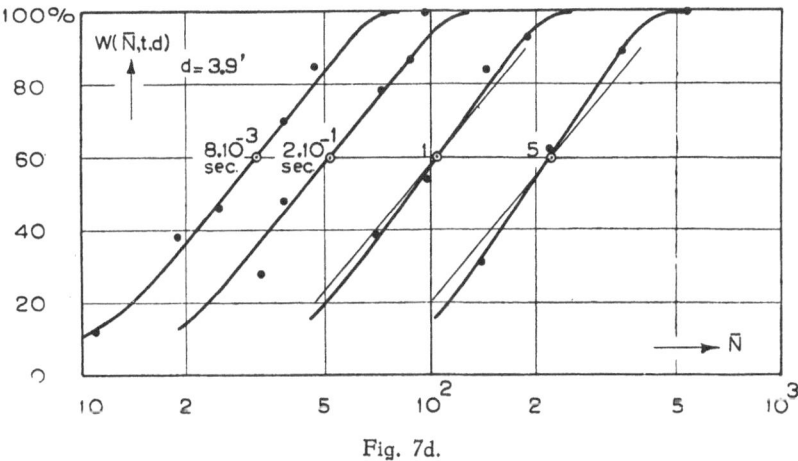

Fig. 7d.

Fig. 7a, b, c, d.
Theoretical and experimental (thick) curve for $W(\bar{N}, t, d)$, the chance
of observation as a function of $\bar{N}$ for different combinations of $d$ and $t$.
The theoretical lines are adapted to the experimental ones by making
them coincide at $W(\bar{N}, t, d)$ 60 % and $t = 8.10^{-3}$ sec.

## 2. The discussion of the deviations.

When $t > 3\ \tau$ together with $d > 3D$ considerable deviations from
the theoretical values are found in the probability and threshold
curves of figures 7 and 8. Also the shape of the curves of figure 7

31

for small times differ somewhat from the, theoretical shape for $k = 2$. The threshold values show a much steeper dependence upon $t$ and $d$ than the two-quanta assumption requires. Such a dependence (proportional with $d^2$ respectively $t$) was already known from earlier publications. As previous authors do not take into account the statistical character of vision, so that the threshold value could not be defined exactly, their measurements are unsuitable for quantitative comparison with ours.

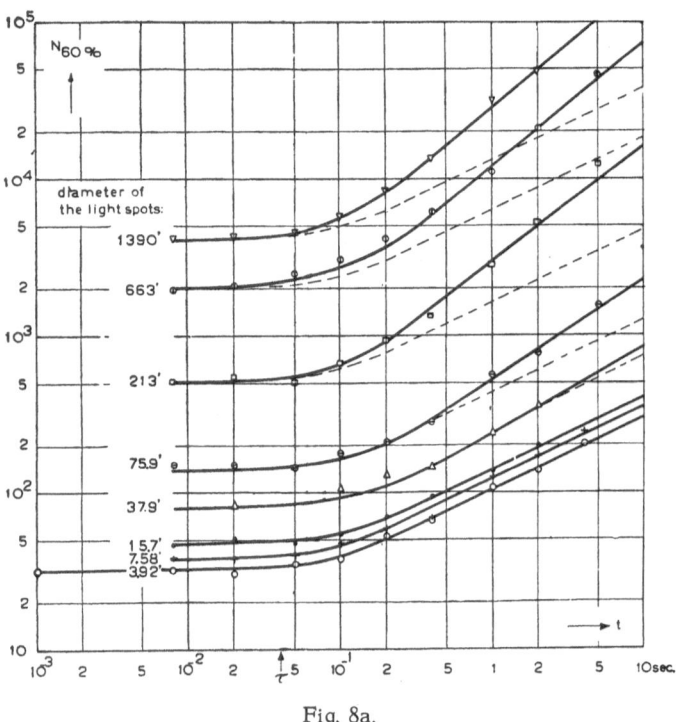

Fig. 8a.

The number of quanta, necessary for a chance of observation of 60 % as a function of $t$, the durvtion of the flash, again both theoretical (dotted curve) as well as experimental (full-drawn curve) for various values of the diameter of the light spot.

Before analyzing these deviations some remarks on the quanta explanation are relevant.

It can be asked whether the behaviour of the threshold-values could also be described by the assumption that the sensitivity of the eye with regards to the absorption power of the photochemical material $f$ and of the number of absorbed quanta $k$ fluctuates with time on every place of the retina and would differ from place to place at every moment.

32

Under optimal conditions $\bar{N}_{60\,\%}$ is about 20 so that the most important numbers $k$ should be smaller than 10, as about half of the number of quanta incident on the eye for the region of wave-lengths employed reaches the retina. Moreover it is very probable that not all quanta are absorbed in the photochemical material of the rods to produce a nerve-impulse that contributes to light-perception. For this reason the most important numbers for $k$ would be very few.

Were a considerable number of light perceptions to arise from the absorption of 3 or more quanta, there must be a rather high possibility for a light-perception

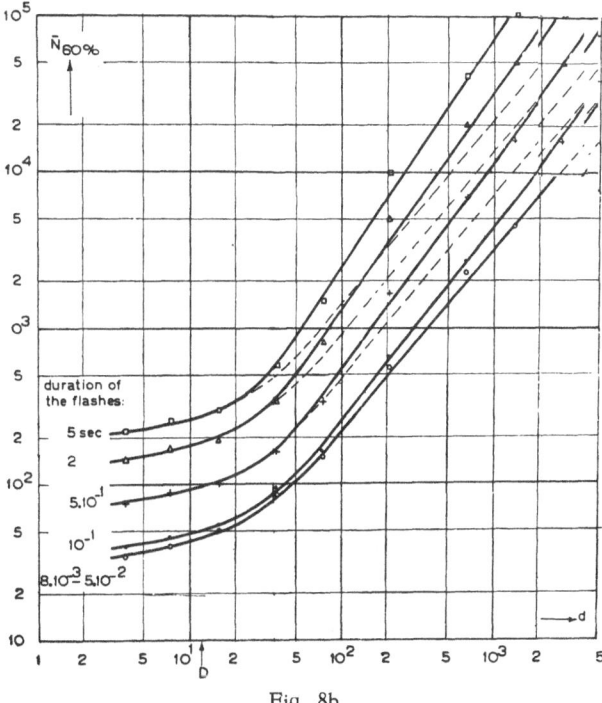

Fig. 8b.

The number of quanta. necessary for a chance of observation of 60 % as a function of the visual angle $d$ of the light spot both theoretical (dotted curve) and experimental (full-drawn curve), for various durations of the flash.

after the absorption of one quantum for the explanation of the obtained slopes of the probability curves when $t$ or $d$ are small.

The slope of the threshold curves as a function of time for small visual angles and as a function of the visual angle for small flashtimes will agree with the lowest possible number of $k$ when $t$ or $d$ is very large. Indeed the slopes of these functions increase very fast with the number $k$ so that the slope belonging to the lowest k will determine the treshold curves.

Moreover there is no evidence to assume that the fraction $f$ fluctuates with time when the eye is completely dark-adapted.

Of course it might be possible that sometimes the nerve-impulses of two rods by which a quantum is absorbed does not succeed in cooperation for a ligth-perception and a third impulse is required for cooperation with one of the first impulses but this does not influence the slope of the threshold-curve with time or visual angle when $t$ or $d$ are large. Anyhow the lowest possible value of $k$ is 2.

Analyzing the possible causes for the deviations from the two quanta explanation in the threshold-values when $t$ and $d$ are large, a number of facts which can not be responsible for the deviations have first to be considered.

$\alpha$.  When $k$ fluctuates with time and differ in place in the manner mentioned the result should not be that deviations occur such as in our experiments. Indeed when $d$ increases the slope of the threshold curve as a function of $t$ will decrease and come in agreement with the smallest possible number of $k$. When $t$ and $d$ are large enough they are indifferent to possible fluctuations in place of $f$.

   The change in slope of the curves mentioned for increasing $t$ respectively $d$ is just in the opposite direction and changes from the slope corresponding to the case $k = 2$ to almost $k = 3$.

$\beta$.  According to Østerberg [8]) the number of rods pro mm² is a function of the place of the retina, sothat $f$ and $D$ can vary in the region of the retina involved in the experiments. As in $\alpha$) it is obvious that this cannot explain the influence of $t$ on the slope of the threshold-curves as a function of $d$.

   It could only lead to the introduction of a correction to the slope of the threshold curve as a function of $d$, independent of the value of $t$. The theoretical curves in Fig. 8 b are not corrected for the variation mentioned, the correction turned out to be small. When $D$ and $f$ are independent of the place, $\overline{N}_{60\%}$ must be exactly proportional to $d$ in Fig. 8 b. When moreover $k$ fluctuates in the manner mentioned under $\alpha$) the slope of the threshold-curves as a function of $d$ will decrease and become in agreement with the smallest possible number of $k$ when $d$ and $t$ are large enough (apart from the small correction for $f$). Again the change in slope of the curves as a function of $d$ for increasing $t$ is in the opposite direction.

$\gamma$.  If variations of $\tau$ in the region of the retina, involved in the experiments, would furnish an explanation for the deviations found, the experimental curve of the threshold as a function of $t$ would lie between the theoretical curves corresponding to the two extreme values $\tau$ found in the region involved. To explain the deviations found these two extremes should differ at least by a factor 50. Measurements of $\tau$ on different places of the retina, gave at most variations of 50 %. For that reason these variations cannot explain our experimental results.

$\delta$.  The variation of the concentration of the visual purple with time cannot play an important role, because, taking for example the case that $d = 100'$ and $t = 5$ sec., only 170 quanta are absorbed by the visual purple. Assuming an average of 100 rods in a sensitive unit in this area with $d = 100'$ the total number of rods is about 6500, so that only 2.6 percent of the rods have absorbed a quantum. During the flash this percentage is still smaller and decreases with increasing diameter and duration of the flash. In view of the fact that the rods cannot influence each other except for their connection by the nerve system, it is not conceivable, that a change in the concentration of the visual purple in one rod, causes such a change in another rod. The deviation mentioned needs more than 60 % for this percentage.

The error in the number of quanta incident on the eye determined in the way mentioned on page 27 may be about 25 percent but is for all strenghts of the current and all distances exactly the same and cannot influence the slope of the curves on which the two-quanta explanation was based. Only the values of $f$ depends on this error. As we did not use monochromatic light the quanta intensity of every wave-leng'h was given a special weight in accordance with the international scotopic luminosity curve. This curve was normalized so as to make its optimal value equal to one; so that $f$ in our experiments refers to the wave-length of the optimal value of the luminosity curve. This wave-length coincides with the op'imal power of absorption of the visual purple.

The actual course of the threshold value with wave-length may differ from the international curve. The way in which the energies mentioned change with the current is practi~ally not influenced by the differences, which may occur in the threshold curve. These only influence the value of $f$. The differences between the real threshold curve and the international curve increase with increasing absorption capacity. When our experiments result in $f$-values of about 10 % or smaller, the influence is negligible. Taking into account the possible error in the absolute level of the energy and in the course of the threshold curve with wave-length for the one observer for whom $f = 28$ percent was found, $\overline{N}$ 60 % under optimal conditions will not exceed 14 quanta. This number is obtained by assuming the eye to be equally sensitive for all wave-lengths, so that it is the real number of quanta incident on the eye.

It is therefore clear that the deviations from the two-quanta theory when both $t$ and $d$ are large might be caused by a phenomenon other than that found till now in the quanta explanation. *The deviations can be explained by the fact that after some time after the beginning of the flash the chance for a perception of light after the absorption of 2 quanta is impeded. This state has developed after a time $T$ (about 0,1 sec.) and extends over an area $O$ (about 20').* It is not impossible that this phenomenon is related to the $\alpha$-adaptation described by Schouten and Ornstein. [23])

We know, that two quanta have to be absorbed by the visual purple within $\tau$ sec. and within an area corresponding to a visual angle $D$, in order to give a light impression when the durations of the flash or the visual angles are small. In a sensitive unit with a visual angle $D$ there are about 100 rods. As the chance is negligible, that both quanta are absorbed in the same rod, it is clear that after the absorption of one quantum there must be already a certain change: an impulse will be transmitted by the nerve connection of a rod when one quantum is absorbed in it. Now one gets an impression of light when a second quantum is absorbed within a visual angle $D$ and within a time $\tau$. Since the only way for interaction between two rods is their nerve connection, after the absorption of one quantum there will be some nerve change extending over a certain area of the retina.

This nerve-change of the apparatus for vision caused by the absorbed quanta can give rise after some time and over a certain area to the mentioned condition of an impeded chance for light-perceptions. It is even possible that this situation is caused by absorbed couples of quanta which have satisfied to the two quanta conditions. It is clear that the change by these couples of quanta is also a nerve change. When couples — and not quanta separately absorbed — give rise to the situation discussed, it is possible that the deviations are of psychological origin. When $d$ and $t$ become large it is possible that the difficulty to draw the attention to the whole area of the test spot during the long time of the flash, has an influence on the threshold values similar to the deviations under discussion.

De Vries, [12]) suggests that the retina grey should be of importance in the region of low brightness. We found that when flashes light up for a test person with an average number of quanta unknown to him, he never has an impression of light when the number of quanta is zero. It seems, therefore, probable that the retina grey does not give the same impression as a normal flash of light.

But for a possible influence in view of the impeded chance it can be of importance when $d$ and $t$ are large.

When the impeded chance is not caused psychologically, the mechanism of transmitting of impulses by the fibers and cells must explain the facts, sothat a physiological aspect is introduced.

A further discussion is given in chapter VIII. When $t < \tau$ or $d < D$ the deviations do not occur and we will use the methods $A$, $B$ and $C$ of chapter I for the study of the properties of the several kind of receptors of the human eye.

# III.  THE TWO-QUANTA HYPOTHESIS AS A GENERAL EXPLANATION FOR THE BEHAVIOUR OF THRESHOLD-VALUES FOR THE SEVERAL RECEPTORS OF THE HUMAN EYE.

This chapter dealt with the experiments on the threshold-values for various wave-lengths for periferal and for foveal vision.

## 1.  The experimental arrangement.

For these measurements it is necessary to use rather good mono-chromatic light. For this purpose we made the experimental arrangement represented in figure 9.

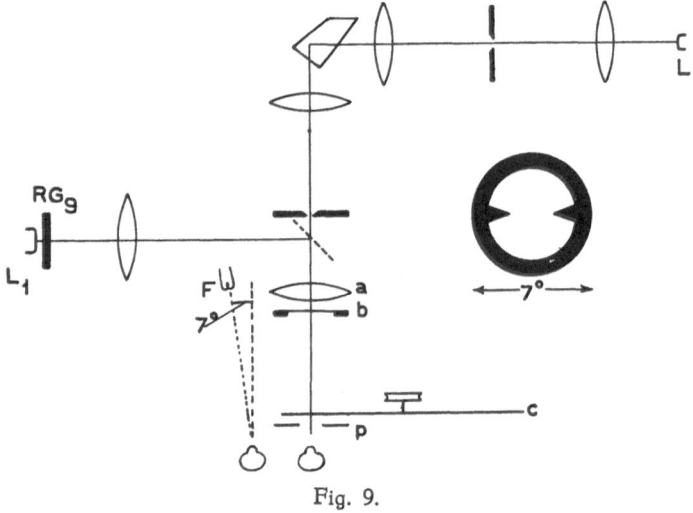

Fig. 9.

The tungsten ribbon filament lamp $L$ is focussed on the slit of a Hilger monochromator by means of a prism of constant deviation. The second slit is focussed by the lens $a$ on the artificial pupil $p$ of 2 mm. diameter. The visual angle could be changed by dia-phragms $b$ in front of the lens $a$. The time of observation could be adjusted by a disk $c$ driven by a synchronous motor. The range of wave-lengths let through by the monochromator never exceeded 150 Å. For the peripheral as well as for the foveal measurements the right eye was used. This eye was fixed peripheral by means of a red fixation-light observed with the fovea of the left eye, in such a way that the spot of the retina with which the flashes were seen was 7° nasal from the fovea of the right eye. For foveal measurements a light source fixed by the left or right eye is of no use as the flashes would then appear on the same place of the visual field as the fixation-light. We used for the foveal measurements a ring concentric with the place of the flashes and provided with two arrows, as shown in fig. 9. This figure was illuminated with white light as weak as possible so that it was just visible to the same eye with which the measurements were performed. The diameter of the ring was about 7 degrees so that its image activated the peripheral rods. The observer tried to fix as well as possible the centre of this figure. In this centre the flashes were seen.

As regards the foveal measurements the fixation figure and the flashes were seen in this way with the same eye so as to prevent deviations in fixations by the non-parallelism of the two eyes. The intensity of the flashes was varied by means of the current of the tungsten ribbon filament lamp. The energy at the pupil $p$ was measured with a thermopile calibrated for absolute value. The eye of the observer was always completely dark adapted before the experiment was started.

## 2. The peripheral threshold-measurements.

In all previous chapters the experimental circumstances were such that only the peripheral rods were stimulated. When we now vary the wave-length from the extreme blue to the extreme red we pass from rod vision to cone vision, so that a region of wave-lengths will exist in which the sensitivity of the two systems is of the same order of magnitude. Outside this region the slope of the probability curve giving the chance of observation as a function of the average number of quanta per flash will on the blue side refer to the rod system and on the red side to the cone system. As already pointed out in Chapter I we can only deduce from this slope the upper limit of the number of quanta $k$ necessary for the two systems. It is possible that the slopes of the probability curves do not agree, even when the two numbers $k$ are equal. Indeed, the time and distance within which the $k$ quanta must be absorbed may differ for the two systems so that the deviations from the theoretical slope according to Poisson may differ for the two systems for reasons mentioned on page 23.

*A question of great importance is whether or not the several systems react independently. In other words: will the absorption of*

$N_1$ quanta by the rod system and $N_2$ quanta by the cone system result in a light-perception or not when both $N_1$ and $N_2$ are smaller than the numbers k required by each system separately.

The shape of the probability curve in the region within which the sensitivities of the two systems are of the same order will depend on these curves for the two systems separately. In the case of independence, so that the stimuli of the systems cannot cooperate to produce a light-impression, in part of this region the slope of the resulting probability curve will be steeper than those of the two curves of the separate systems. When the two systems are equal sensitive the resulting chance is for the case of independence

$$W(\overline{N}) = 1 - (1 + f\overline{N})^2 \, e^{-2f\overline{N}}$$

This function increases faster with $\overline{N}$ compared with the corresponding function for the case of dependence

$$W(\overline{N}) = 1 - (1 + 2f.\overline{N}) \, e^{-2f.\overline{N}}$$

Fig. 10 shows the probability curves for various combinations of flashtime and visual area, related to four wave-lengths, the dark-adapted right eye, 7° nasal from the fovea. Besides the slight increase of the slope with increasing time or visual angle which was to be expected it appears that the curves are steeper in the red end of the spectrum.

In one of the graphs of fig. 10 we plotted also the theoretical slopes according to Poisson's law. We can deduce along the lines given in chapter I, that for rod vision the upper limit of the number of effectively absorbed light-quanta k necessary for the light-perception is 2. From the previous chapters we know that k is 2. For cone vision this upper limit is obviously higher, namely about 3. As we can expect that at about 6000 Å the rods and cones are almost equally sensitive, the slopes of the curves in this region can give information about the mutual dependence of the different systems. It is noticed that the slopes are not steeper compared with the slope for 7000 Å and 5100 Å. If they were independent the slope should be steeper compared with the slope for 7000 Å as well as for 5100 Å. Although the reliability of the measurements is rather poor for drawing this conclusion, it is likely that the systems are not independent at the threshold of vision.

In section 4 we shall use a much more appropriate method by which this conclusion is confirmed.

For the exact determination of $k$ we must make use of the other methods indicated.

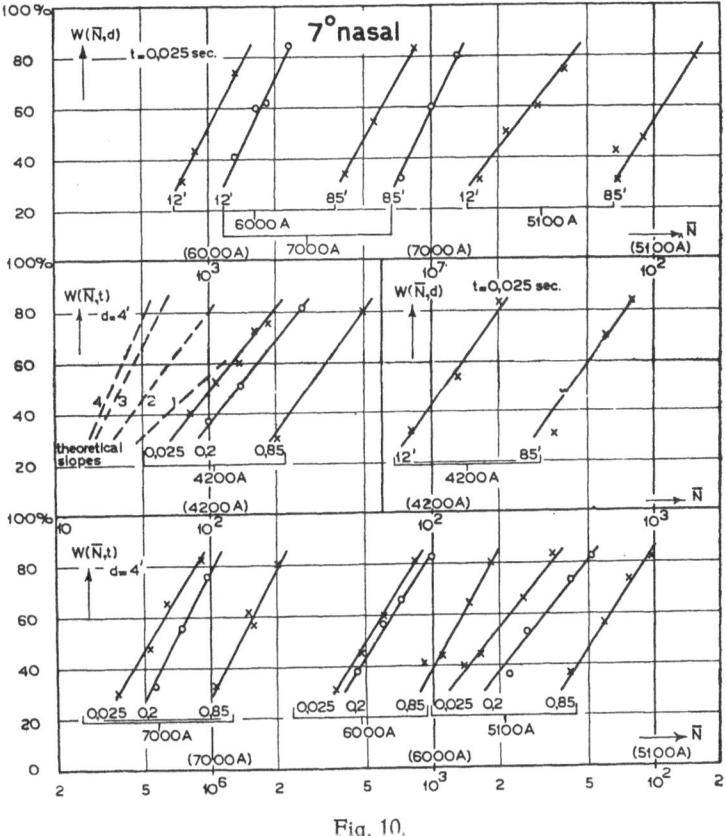

Fig. 10.

The chance of observation $W(\bar{N}, t, d)$ as a function of $\bar{N}$ for various wave-lengths and for different combinations of $d$ and $t$. The theoretical slopes for the 1, 2, 3 and 4 quanta case according Poissons law are submitted (dotted lines). The level of energy for each wave-length is indicated at the abcis.

Fig. 11a and b shows the threshold-values $\bar{N}_{60\%}$ for various combinations of flash-time and visual angle. Regarding the curves as a function of time of fig. 11a we see that the shape is almost the same for all wave-lengths. The slope for $t$ is very large agrees for all wave-lengths with $k = 2$ as for this case $\bar{N}_{60\%}$ has to be proportional to $t^{1/2}$. It is impossible to cover them with $k = 3$ as proportionality with $t^{2/3}$ would be required.

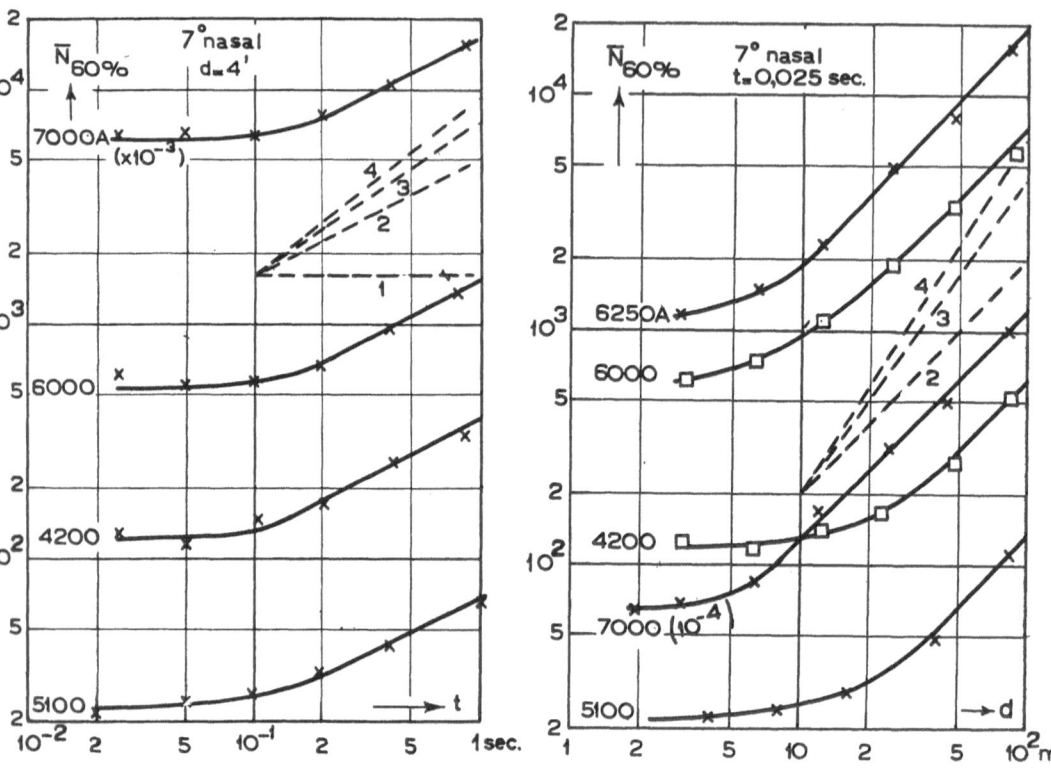

Fig. 11a and 11b.

The average number of quanta, necessary for a chance of observation of 60 % as a function of $t$ respectively $d$
for various wave-lengths.

The curves of fig. 11a do not show the existence of the two
separate systems because the differences in the time-value at which
the $\overline{N}_{60\%}$ from being independent of $t$ alters into a function of $t$,
proportional to $t^{1/2}$, is very small and might be due to the unavoi-
dably limited accuracy of measurements of this kind.

*We can conclude that for the rod and cone system the number k
is 2 and these two quanta must be absorbed in a time, which is the
same for the two systems, and is equal to about 0,04 sec.*

The slope of the threshold-curves $\overline{N}_{60\%}$ as a function of the visual
angle $d$ of the flashes when $d$ is large agrees again for all wave-
lengths with $k = 2$, since $\overline{N}_{60\%}$ is proportional to $d$, as required for the
two quanta case. Is is impossible to cover them with $k = 3$ as proportionality with $d^{4/3}$ would be required. Using the method of the pro-

bability curves (see fig. 10) we found for 6000 Å and 7000 Å an upper limit for $k$ equal to 3 and for smaller wave-lengths 2. This confirms the fact that in the region of the longer wave-lengths another kind of receptors is stimulated than in the green and blue part of the spectrum. This is further confirmed by the appearance of a sensation of colour in our threshold measurements in this region of the spectrum. For all wave-lengths larger than 6000 Å we experience a reddish sensation, for the other wave-lengths a colourless rod-sensation. In the behaviour of the threshold-curves as a function of the visual angle we observe a well-defined difference between the rod- and the cone- system. The visual angle value at which $\overline{N}_{60\,\%}$ from being nearly independent of $d$, becomes a quantity proportional to $d$, is almost equal to the distance $D$ within which the two quanta $k$ have to be absorbed in order to result in a light-impression. It is seen that for the wave-lengths 6000 Å and 7000 Å this value becomes about 3 to 4 minutes whereas for the other wave-lengths $D$ is about 12 minutes. Such a dependence on the wave-length of the distance $D$ within which the two quanta have to be absorbed is only possible when there are more than one system.

In the region of wave-lengths within which the sensitivity of the two systems is of the same order of magnitude the threshold-curve as a function of $d$ will change with wave-length from the curve belonging to the rods of the curve of the cones. The way in which the threshold-curve is changed from being independent of $d$ to being proportional to $d$ depends not only on this ratio, but also on any possible interaction between the two systems and rectangularity of the chance for a lightperception as a function of the actual distance between the two absorbed quanta $W\,(\triangle T,\,\triangle d)$. For this reason it can not be ascertained from the way in which the threshold-curve is changed from $D \infty 12'$ to $D \infty 3'$ whether or not the two systems can cooperate, nor the ratio between their sensitivities as a function of wave-length can be determined out of this change.

Accepting the results of Østerberg [8]) there are about 100 rodlike receptors in an area with a visual angle of 10 minutes 7° nasal from the fovea. As the chance is negligible that at the threshold for rod vision the two quanta are absorbed in one of this number *a rod reacts on the absorption of one quantum. The rod will send a nerve impulse to its nerve-connection. Two impulses of rods within the distance D mentioned and within $\tau$ seconds will cooperate and give rise to a light-perception.*

In a region of 3 to 4 minutes within which the two quanta must be absorbed for cone vision in the extreme red, the number of cone-like receptors is very small at 7° nasal from the fovea namely about 2 according to Østerberg [8]).

*So when we identify the rod-like and cone-like receptors of Østerberg with our rod respectively cone-system we must conclude that for the cone-system 7° nasal as far as the wave-length is larger than about 6500 Å the two quanta have to be absorbed within the selfsame receptor within τ seconds.*

This may be the reason why the probability-curves of fig. 10 are steeper in the red compared with the slopes in the blue part of the spectrum, in agreement with the observation in chapter I section 2. Of course it might be that the nerve fibers of the separate cones are not long enough to reach the nerve-connections of the next cone as the number of cones in this region of the retina is so few. It is clear, that in this case impulses from different cones can not cooperate.

The time within which the two quanta must be absorbed does not differ for the two systems.

As for the rod-system each of the two quanta gives rise to a nerve-impulse which mutually cooperate in the nerve-system, *it is very probable that each of the absorbed quanta in a cone for 6500 Å or larger also causes a nerve-impulse and that these two impulses cooperate in a nerve-element of the retina similar to the element in which the rod cooperation occurs.* If the two quanta absorbed in a cone would give rise to one single nerve-impulse it is rather strange that the time within which the two quanta must be absorbed in the cone agrees with the time mentioned for the rods which is of nervous origin.

From the experiments of section 4 it is comfirmed that from each absorbed quantum arises a nerve-impulse.

### 3. The central-foveal threshold-measurements.

Fig. 12 shows probability curves for various combinations of visual angle and flash-time for the central-fovea of the dark-adapted right eye. Neglecting the possibility that for one or more of the chosen wave-lengths the sensitivities of two or more receptor-systems are of the same order of magnitude and moreover react independently we can deduce from these curves for the smallest $t$ and $d$ that for all wave-lengths the upper limit of the number $k$ is 3.

Most of the curves are steeper than the Poisson formula for $k = 2$ requires and less steeper than for $k = 3$. In section 4 we shall give the experiments as to whether or not an interdependence between various cone-systems of the fovea exists.

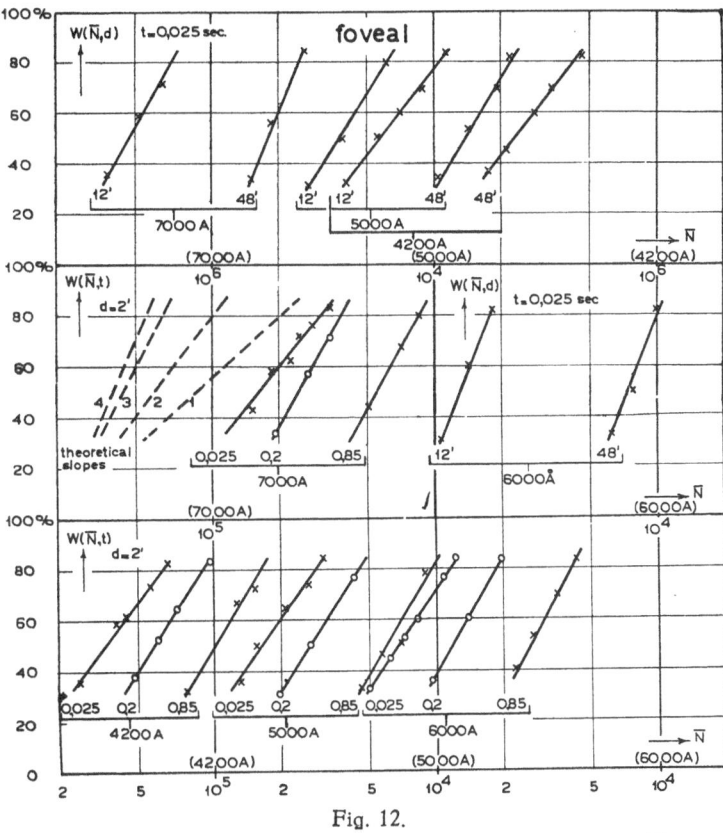

Fig. 12.

The chance of observation $W(\overline{N}, t, d)$ as a function of $\overline{N}$ for various wave-lengths and for different combinations of $d$ and $t$. The theoretical slopes for the 1, 2, 3 and 4 quanta case according Poissons law are submitted (do'ted lines). The level of energy for each wave-length is indicated at the abcis.

For the exact determination of the number $k$ we again used the two other methods. In fig. 13 we give the threshold-values $\overline{N}_{60\,\%}$ for various wave-lengths as a function of time and visual angle. *The shape of the curves as a function of time is again almost the same for all wave-lengths. The slope with time when $t$ is large agrees with the case $k = 2$.* From these curves the existence of more systems is not apparent as here too the difference in the time-value, at which $\overline{N}_{60\,\%}$ shifts from being independent of $t$ to the region in which $\overline{N}_{60\,\%}$ is proportional to $t^{1/2}$, is very small.

*The slope of the threshold-curves $\overline{N}_{60\,\%}$ as a function of the visual angle $d$ of the flashes when $d$ is large agrees again for all*

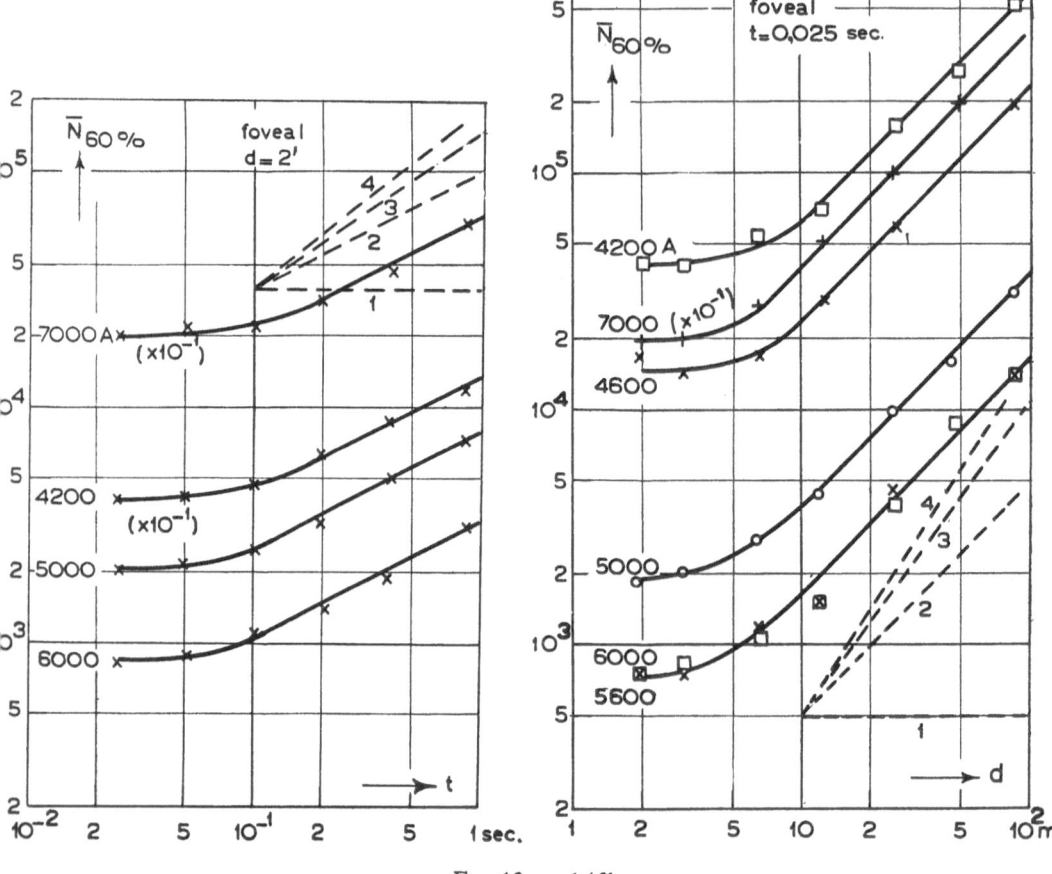

Fig. 13a and 13b.

The average number of quanta, necessary for a chance of
observation of 60 % as a function of $t$ respectively
$d$ for various wavelengths.

*wave-lengths with* $k = 2$. Till now we have from our measurements
no evidence that for the foveal vision more than one receptor-
system exists. The slopes of the probability curves as well as the
threshold-curves as a function of time or visual angle do not give
any clear indication of the actual existence of more than one system.
For the measurements of the dependence of $\bar{N}_{60\%}$ on the visual
angle it is of the greatest importance to apply with the utmost care
the necessary corrections for the chromatic aberration of the eye.
Without these a great dependence on the wave-length of the region
of the visual angle $d$ in which $\bar{N}_{60\%}$ is almost independent of $d$

is found as without a good correction the illuminated area of the retina will not decrease below a definite value, dependent on the wave-length, when the visual angle of the flash is decreased. In our experimental arrangement it was necessary for the right eye of the observer to apply the above correction with a lens of -0,75 dioptries for 7000 Å and with -2,25 for 4200 Å in front of the artificial pupil.

*Summarizing: We found that for foveal vision again two quanta must be effectively absorbed within a distance of about 2 to 4 minutes and within a time of about 0,04 Sec. This time agrees once more with the time found for the peripheral cone- and rod-system.*

*Of course it is necessary to measure this time for more places of the retina but it seemed that the nerve-element in which the co-operation of the two impulses of the two absorbed quanta occur is for all receptor-systems for the whole retina of the same kind.*

There are several reasons for the impossibility to conclude from our foveal measurements that for foveal vision the two quanta must be absorbed in the same receptor. In a region of about 2 to 4 minutes there are according to Østerberg [8]), 20 to 90 receptors in the central fovea. The chance that the two quanta are absorbed in two different receptors is not negligible, even down to the smallest areas dealt with. The area of the retina covered by the flash will be enlarged by the size of the Airy diffraction disc. For the experimental arrangement described in section 1 we must add to the size of the flashes 2,3 to 1,3 minutes caused by diffraction. In order to test whether this influences the region of $d$ in which $\overline{N}_{60\%}$ is almost independent of $d$ we made an arrangement for which the diffraction was smaller, and only about 0,8 to 0,4 minutes had to be added to the size of the flashes computed purely geometrically. We found no influence on the region of $d$ within which $\overline{N}_{60\%}$ is almost independent.

When the various kinds of receptors are not evenly distributed over the fovea as suggested by Hartridge [24]) but form clusters of cones of the same kind the distance $D$ seemed to be larger. When the area of the retina covered by the flash is smaller than such a cluster a part of the flashes for a certain wave-length can fall on receptors which are insensitive to the wave-length employed. For this reason the smallest angles $\overline{N}_{60\%}$ would become higher than agrees wit a homogeneous distribution.

There might be another reason by which the region of $d$ in

which $\overline{N}_{60\%}$ is almost independent of $d$ can be influenced, namely the possibly small involuntary movements of the eye which always occur even when the observer tries to fix his eye as well as possible. By these movements the absorption of two quanta within a certain area is impeded and increases the threshold when the visual angle of the flash is of the same order of magnitude as the size of such an area. But if we accept the results of Østerberg, for peripheral cone vision in the red region the two quanta must probably be absorbed in one cone. So it is possible that in the fovea too this condition must be satisfied for these wave-lengths in the red. In the previous paragraph we also suggested that each quantum absorbed in a cone 7° nasal in the red region may give rise to a nerve-impulse.

These two nerve-impulses must be transmitted by the nerve-connection of this cone within a time of about 0.04 sec. for the light perception. From the electrophysiology of nerve fibres it is known that the impulses can only follow each other with intervals greater than the refractory period. These periods are about $10^{-4}$ seconds. It would seem therefore that with decreasing flash time the $\overline{N}_{60\%}$ will increase for values of $t$ of $10^{-4}$ seconds. As far as we know, such an effect has never been found. In our opinion, this is not an objection to the suggestion that under certain circumstances (for instance in the red region of foveal and of peripheral vision) two quanta must be absorbed in one cone for the light perception and that the two quanta each cause a nerve-impulse in the nerve-connection of the cone. The way in which the absorption of a quantum give rise to a nerve impulse of the cone is practically unknown. In any case the time between the absorption of a quantum and the start of the impulse in the nerve-connection will not be infinitely small, and will differ time after time by the inevitable natural fluctuations in the course of the process transmitting the energy of the quantum to the starting of the nerve-impulse. If this fluctuation in time is longer than the refractory period of the nerve-connection two quanta absorbed at the same moment will give rise to two separate nerve-impulses in the nerve-fiber.

At the Physiological Conference in Oxford and the International Conference on Colour Vision in July 1947 Hecht dealt with among other things his experiments concerning the determination of the number $k$ for the absolute and differential threshold for cone vision. His conclusion that for the absolute as well as for the differential threshold the absorption of 4 quanta in a cone of the test-spot is necessary, is founded on probability-curves obtained by the first method described in chapter I. As we pointed out at the time this method is not suitable for the exact determination of $k$, moreover the case $k = 4$ can certainly not explain the dependence of the threshold with time or visual angle. As we found $k = 2$ for the absolute threshold we have also serious objections to the conclusion as regards the differential threshold (intensity discrimination). It is hard to understand how for all intensity levels of the differential threshold the behaviour will be in agreement with $k = 4$ except when the surrounding intensity is zero. Our preliminary experiments on the intensity-discrimination are in agreement with the suggestion of de Vries [12]) excepting for the lowest intensities.

According to de Vries the intensity discrimination is proportional to the reciprocal of the root square of the intensity.

## 4. The mutual dependence of the receptors of different kinds in the retina.

Pirenne [20]) used a method for investigating any possible mutual influence at the threshold of vision between the two eyes. We used this method in a somewhat different way for investigating whether a mutual dependence exists between the receptors of different kind in the same eye.

Suppose the chance of observation of a flash presented separately with an average number of quanta $\overline{N}_1$ is $W(\overline{N}_1)$ and of another flash $(W(\overline{N}_2)$. When these flashes are presented simultaneously to the observer the resulting chance of observation will be $W(\overline{N}_1, \overline{N}_2) = 1-[1-W(\overline{N}_1)][1-W(\overline{N}_2)]$ for the case that the receptor-systems stimulated by the two separate flashes react completely independently. *When we found a greater resulting chance, we can conclude to dependence.*

For the dark-adapted eye at $7°$ nasal from the fovea at the threshold of vision only the cones are stimulated by light of wavelength 7000 Å and only the rods by quanta from about 6000 Å downward. In the experimental arrangement shown in fig. 9 a glass plate was inserted behind the second slit of the monochromator. By means of this plate the light of a second tungsten ribbon filament lamp $L_1$ is focussed by lens a on the artifical pupil $p$. The light of $L_1$ first passed a Schott RG 9 filter so that only quanta of 6900 Å and greater wave-lengths are used. Now we first determined the current value of L for which the chance of observation of the flash is about 30 %. The visual angle was always 25' and the time 0,025 seconds. We next determined the current values of $L$ for which the chance of observation of flashes of various wave-lengths was again about 30%.

By the flashes of $L_1$ only the cones are stimulated and by the flashes of $L$ only the rods. We now expose the eye to the flash of $L_1$ together with the flashes via the monochromator.

The result is that on the same spot of the retina and at the same time quanta are absorbed of two kinds, namely of 6900 Å and of another wave-length depending on the adjustment of the mono-chromator.

In table I we give the results of these experiments. In the first two columns the chances are given for the components of the mixture separately, in the third the measured resulting chance. From the first and second column the chance is computed for the case of

Table I. Chance of observation of flashes with visual angle of 25' and time 0.025 second 7° nasal from the fovea for various wave-lengths and combinations of wave-lengths.

| W ($\overline{N_1}$) | W ($\overline{N_2}$) | W $(\overline{N_1, N_2})$ experimental | W $(\overline{N_1, N_2})$ theoretical independent | W $(\overline{N_1, N_2})$ theoretical dependent |
|---|---|---|---|---|
| 6900Å, 27 % | 5320 Å, 43 % | 85% | 58% | 72% |
| 6900Å, 27 % | 5320·Å, 30 % | 80% | 49% | 62% |
| 6900Å, 27 % | 4800 Å, 25 % | 64% | 45% | 60% |
| 6900Å, 27 % | 4620 Å, 25 % | 67% | 45% | 60% |
| 6900Å, 27 % | 4400 Å, 25 % | 67% | 45% | 60% |
| 6900Å, 27 % | 4200 Å, 30 % | 65% | 49% | 62% |
| 6900Å, 27 % | 4700 Å, 25 % | 72% | 45% | 60% |
| 6900Å, 27 % | 5060 Å, 27 % | 75% | 47% | 72% |
| 6900Å, 27 % | 3900 Å, 30 % | 77% | 49% | 62% |
| 6900Å, 27 % | 4500 Å, 22 % | 64% | 43% | 57% |
| 6900Å, 27 % | 4500 Å, 40 % | 80% | 56% | 70% |

complete independence and these values are given in column four. In the last column the resulting chances are given for the case that the behaviour would have been as of one system and on the assumption that the slope of the probability curve agrees with the Poisson formula for the two quanta case.

In order to make sure that the two flashes coincide the visual angle of the flashes is chosen rather large *) and the slope of the probability curve for the separate flashes will be steeper than agrees with the smallest visual angle (see chapter I section 2). The slope will certainly be steeper than the slope of the purely theoretical Poisson curve, so that in the case that there exist mutual dependence of the systems the resulting chances will even be greater than the chance of the last column.

*From these considerations we can conclude that the rods and cones react almost completely mutual dependent, as the measured resulting chances are indeed greater than the chances of the last column.*

From the preceding paragraphs and chapters it is clear that a rod reacts on the absorption of one quantum and that two rod impulses result in a light-perception when the two quanta conditions are satisfied. For peripheral vision in the red region of the spectrum it is probable that two quanta must be absorbed within one cone.

*From the interdependence of the rod- and cone-systems in the periphery we must conclude that if one quantum is absorbed in a cone which is sensitive in the red region of the spectrum and one other quantum is absorbed in a rod situated within a distance D' of the cone a light impression will result so that certainly each absorbed quantum in a cone result in a separate nerve-impulse.*

In table II we give the results for the foveal measurements. *The measured resulting chances show, that the receptor system sensitive for certain wave-length reacts in any case in complete interdependence with the systems sensitive for an other wave-length.* From what is known concerning the fundamental response curves, reviewed by Wright [25]), the dependence found, cannot be explained by a possible overlapping of the sensitivity curves of the three systems. When the two systems are completely independently

---

*) We were forced to choose the visual angle rather large namely 25'. It is difficult to check whether for peripheral vision the quanta of the different wave-lengths fall indeed on the same spot of the retina when the flashes are smaller. When for foveal vision the two flashes coincide as regards place, the quanta of the two wave-lengths, if they differ much, are not focussed in the same way on the retina when the eye is turned to 7° nasal as the chromatic deviations of the eye are rather large.

Table II. Chance of observation of flashes with visual angle of 25' and time 0.025 second from foveal measurements for various wave-lengths and combinations of wave-lengths.

| $W(\overline{N_1})$ | $W(\overline{N_2})$ | $W(\overline{N_1}, \overline{N_2})$ experimental | $W(\overline{N_1}, \overline{N_2})$ theoretical independent | $W(\overline{N_1}, \overline{N_2})$ theoretical dependent |
|---|---|---|---|---|
| 6900Å, 32 % | 4200 Å, 29 % | 73% | 52% | 65% |
| 6900Å, 38 % | 4200 Å, 25 % | 71% | 54% | 66% |
| 6900Å, 38 % | 5250 Å, 28 % | 90% | 55% | 70% |
| 6900Å, 45 % | 4000 Å, 30 % | 88% | 62% | 74% |
| 6900Å, 45 % | 5100 Å, 30 % | 85% | 62% | 74% |
| 6900Å, 45 % | 6000 Å, 25 % | 85% | 59% | 72% |

reacting at the threshold of vision, the resulting chances measured by us should have been smaller for most of the chosen combinations of wave-lengths, even for the curves of Hecht. If our measurements were to be explained by a complete independence, the three-response curves must lie closer to each other than Hechts curves and this is impossible as is proved in chapter V. We must conclude that an almost complete dependence exists between the different cone-systems at the threshold of vision.

*The information on rod- and cone-vision obtained from our threshold measurements and investigations on their dependence, may thus be summarized:*

*A rod reacts with a nerve-impulse in its nerve-connection after the effective absorption of one quantum. A light impression is caused by the absorption of a second quantum in a rod within a distance D of the first and within a time $\tau$.*

*A "red cone" in the periphery gives rise to a light-impression when within a time $\tau$ two quanta are absorbed in it. The mutual dependence of the peripheral rod- and "red cone" systems can only be explained by each absorbed quantum in a cone giving rise to a nerve-impulse in the nerve-connection of the cone and by one rod- stimulus cooperating with one cone-stimulus to produce a light-impression. It is very probable that also for this kind of cooperation the quanta must be absorbed within the time $\tau$ and the distance D. For the foveal cone systems for every wave-length a ligth-impression is caused by the absorption of two quanta within a time $\tau$ and a distance $D_1$ of 2-4 minutes.*

52

*The foveal cone-systems sensitive for a certain wave-length react at the threshold of vision in complete mutual dependence with the systems sensitive for another wavelength sothat*

*a cone of certain kind reacts with a nerve-impulse to his nerve-connection after the effective absorption of one quantum. A light-impression is caused by the absorption of a second quantum within $\tau$ in another cone which must lie within a rather small distance from the first one.*

*A rod-stimulus 7° nasal from the fovea can cooperate with any other second stimulus within a time $\tau$ and a distance $D$ to a light-impression.*

It might be possible that the colour sensation of a light perception of a "mixed stimulus" can differ definitely from the sensations of the stimuli of the separate systems. In the region within which the sensitivity of two of the systems is of the same order of magnitude and very small areas of the retina are illuminated at the threshold of vision, statistical fluctuations will occur in the colour of the light-perception as the colour will depend on the kind of receptor in which the two quanta are absorbed. We have started experiments on this subject together with ten Doesschate. It seems likely that for certain wave-lengths short and small flashes can give rise to more than three discrete colour sensations. (See also reference 24).

Thus a certain kind of flash for instance is sometimes called yellow, sometimes blue or white or green.

## IV.  SOME PROPERTIES OF THE NERVOUS SYSTEM
## OF THE VISUAL SENSE.

### 1.  The nerve elements.

The *neurons* are the conducting elements of the entire nervous system. (See Howell [26]) ). They exist in a wealth of shapes and sizes, yet they all have many features in common: about the nucleus is an accumulation of plasm which is called the *perikaryon* or *cell-body*. From the perikaryon parts of two kinds are given off: *dendrite* and *axon*.

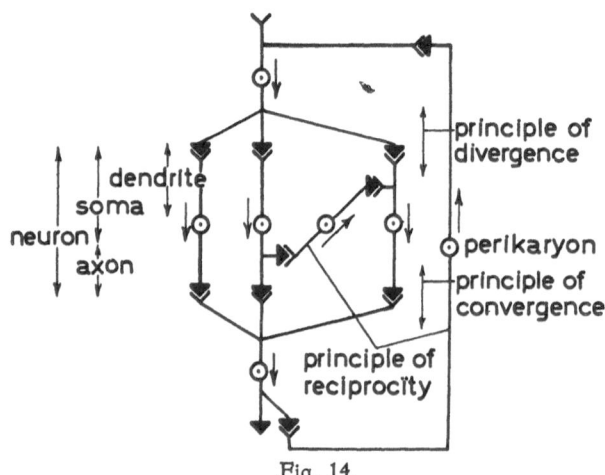

Fig. 14.

Scheme of nerve-elements.

The perikaryon and dendrites are the receiving part of the neuron and are called *soma*.

The normal conduction in a neuron takes place in the direction from soma to axon. The axon can have a number of branches. The actual ending of each branch is a microscopic *disk, bouton* or *pied terminal*, applied to the surface of the receiving part of the contiguous neuron, without protaplasmic connection. The surface of the soma under a pied is a *synapse* and this synapse is the source of the characteristic features of the actions of the nervous system.

The collection of synapses which forms a part of the membrane of the neuron is called *the synaptic scale*.

The retina is a part of the central nervous system. The neurons of the retina are distributed in different layers. These layers consist

54

of three strata of densely packed perikaryons and two intervening synaptic layers consisting of intertwining dendritic and axonic brushes (fig. 15). The structural pattern underlying transmission

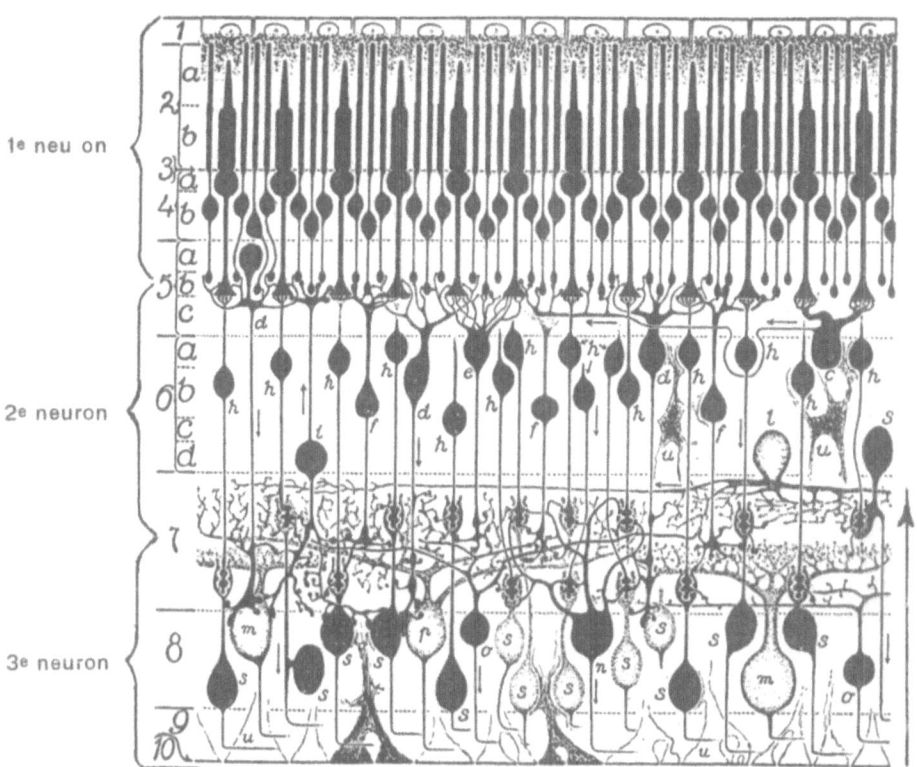

Fig. 15.

Reconstruction of the retina showing the principal neuron types and their synaptic relations. 1. pigment epithelium; 2a, outer segment of rods and cones; 2b, inner segment of rods and cones; 3, outer limiting membrane; 4, outer nuclear layer; 5, outer plexiform layer; 6, inner nuclear layer; 7, inner plexiform layers; 8, ganglion cells (origin of optic nerve fibers); 9, layer of optic nerve fibers; 10, inner limiting membrane. Various types of cells are: c, horizontal cells; d, e, f, diffuse or polysynaptic bipolar cells; h, individual cone (midget) bipolar cell; i, l, amacrine cells; m, n, o, p, r, s, ganglion cells of which s is the individual or monosynaptic ganglion cell. (From Polyak, The retina, 1941).

assumes almost indescribable complexity. Each axon divides many times to supply a number of contiguous somas: *the principle of divergence* (cell c in fig. 15).

55

The neurons in layer 8 are called *ganglion cells*. From these the axons go to the brain. The *nervus opticus* is a bundle of these axons. Each ganglion cell is supplied by a number of axons: *The principle of convergence* (cell n). Besides these principles all of the neurons of the central nervous system are reciprocally connected by many pathways of a greater or lesser degree of complexity: *the principle of reciprocity* characterized by cells conductung in horizontal and backwards direction (cells l, i).

Polyak [7]) made a very extensive investigation of the retina with the Golgi technique.

By a colouring substance occasional neurons are fully impregnated so that the cell body and the dendritic and axonic ramifications of single neurons can be made out. So with this method only the presence of nerve elements can be found but nothing can be concluded about the absence of elements of a certain kind.

Polyak recognized two types of bipolar cells. The first type, which is the most common, is termed the rod and cone (cells d, e, f) bipolars. They have a wide spread of dendritic branches whereby they receive impulses from a group, sometimes large, of rods and cone neurons. The dendritic ramifications overlap, so that a single receptor neuron connects with more than one bipolar cell.

*This observation means that rod and cone systems are incompletely separated in the pathways to the brain. This finding obviously embarrasses the theory of specific receptors and specific nerve energies. It makes understandable and gives an explanation for the dependence between the rod and cone system established by us by experiments on the behaviour of the threshold values described in chapter III section 4.*

The second main type of bipolar cell is the cone bipolar cell (cell h).

Rods never connect with these bipolars; the cone bipolars are related only to the cones and in the fovea centralis each bipolar is connected with only one cone.

The ganglion cells fall into two broadly similar categories: 1) diffuse ganglion cells which connect with a great number of bipolar cells and exhibit to a considerable degree the phenomenon of partially shifted overlap; 2) the individual ganglion cells which are via bipolars connected with one or two cones (in the fovea always with one cone).

*So in the retina two systems of neurons occur. The first is*

*according to Polyak's data exclusively identified with cones.*
*Each cone of this system has at his disposal a private path in the*
*optic nerve. The second system is a mixed rod-cone system and*
*is marked by the convergence, divergence and reciprocity of rods,*
*cones, bipolars and ganglion cells.* An arrangement of this sort
affords a basis for interaction of one retinal area with another
resulting in facilitation and inhibition phenomena. Moreover, the
possibility of interaction is increased by a system of intraretinal
association neurons: horizontal cells, centrifugal bipolars and possibly some of the amacrine cells (See fig. 15) (cells c, i. l).

The axons of these neurons run horizontally for long distances
in the outer plexiform layer.

We found dependence in the behaviour of the cones in the central
fovea (described in chapter III). This seems to be in contradiction with the seperated conducting system found in Polyak's
preparations. We found in the fovea that the impulse of a quantum
absorbed in one cone can cooperate with such an impulse of another
cone situated within about 3' of the first.

*By this reason in our opinion there exist mutual connections*
*between the cones, somewhere in the visual pathways.* We mentioned already that with the methods used by Polyak the absence of
nerve elements of a certain kind cannot be ascertained, so that it
might be that the mutual connections mentioned are situated after
all in the retina. When not, they must be found in the other parts
of the nervous system of the visual sense organ.

Out of the measurements 7° nasal from the fovea it proved that
probably no cooperation occur between the separate cones sensitiv
in the extreme red of the spectrum. *Obviously, this is not due to*
*the fact that these cones should have separately nerve conduction*
*systems.*

Indeed, there exist mutual dependence with the neighbouring
rods: one quantum absorbed in a cone and one quantum in a rod
cause a light impression in agreement with Polyak's mixed rod
cone systems.

By this reason, the absence of cooperation between cones is due
to the small number in this area of the retina: the impulses of
two cones cannot interact as the distance between the cones is
larger than the optimal distance within which their neurons can
activate the retina in horizontal direction. Contrary, the measurement of the distances within which the two quanta must be absorbed

in the retina in order to result in a light impression gives indication about this optimal distance.

## 2. Some properties of the transmission in the nervous system.

The prime function of neurons is the conduction of impulses.

The excitation of sensory end-organs as for instance rods and cones is conducted with impulses in the neurons. The nerve impulse is always associated with an electrical change of characteristic time course, *the action potential*. The only modification of the neuron activity is by alteration of the frequency of the impulses. The weakest response of the neuron is the conduction of one impulse.

The size and velocity of the separate impulses in a neuron is independent of the strength of the stimulus.

This constitutes the *"all or none law"*, a fundamental rule in the activity of nerve fibers.

When a nerve impulse has travelled along a nerve fiber, a certain period of recovery must elapse before another impulse can pass, *the refractory period*. The total duration of the impulse is about $10^{-3}$ seconds, and about $5 \times 10^{-3}$ seconds are needed for complete recovery. The velocity of transmission of the impulse is proportional to the reciprocal of the diameter and ranges from about 120 to 1 meter/sec. The size of impulse increases with increasing diameter.

*When the stimulus to a neuron is weakened to a value below its threshold, no impulses are setted up, but two or more of such stimuli in quick succession are able to excite the neuron.*

This effect is due to the local temporal disturbance of excitability instituted by a subliminal stimulus (vide Howell page 20). This local change is called the *local excitatory state* what's amount is proportional to the strength of the stimulus and is not conducted along the neuron. It extends a short distance along the fiber with an exponential decrement.

To accomplish the performances of the body, transmission of excitation from one neuron to another must take place.

This transmission occur via the synapses: synaptic transmission.

When a single impulse originating from a receptor organ should always be conducted from neuron to all contiguous neurons via the interlocated synapses no neuron of the body should escape from excitation as by the principles of convergence, divergence and reciprocity each neuron is in some way connected with each other neuron of the body. By this reason the transmission via a synapse must be conditional.

*Sherrington* [27]) *found in studying simple reflexes, that it is always possible to make a stimulus so weak that it sets up several centripetal impulses, but yet fails to elicit a reflect response. Therefore it follows that summation of the effects of several impulses is necessary to set up a reflex discharge. Sherrington and co-workers developped that mostly two impulses in quick succession succeeded in arising a reflex (vide Creed page 34).*

*Eccles* [28]) measured the fluctuations of the potential of synapses. It proved that a local and temporal excitation arises on the cell surface on the synapse, when an impulse in the previous neuron reaches the synapse. This effect is called *the central excitatory state*. It has a sharp spatial decrement.

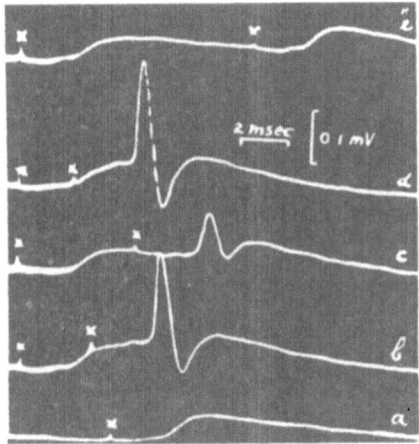

Fig. 16.

Records from 8th ventral root of a cat to two excitations at various intervals (See $\times$ in figure). In *a* and *e* only *a* synaptic potential is setted up, whereas in *b*, *c* and *d* a spike discharge due to facilitation of the two impulses is shown.
(From Eccles. J. Neurophysiol. 9, 94 (1946) ).

A second impulse within a short time interval after the first can cause an increase of the central excitatory state, so that the critical discharge potential is reached and an impulse is transmitted in the neuron.

The critical discharge potential can be reached by spatial summation or by temporal summation.

There are two ways for making this occur:

1e.  *The two impulses are conducted in the same axon.*

As in this case the two impulses can not follow each other within a time interval smaller than the refractory period of the fiber it is necessary that the central excitatory state caused by the first impulse lasts at least for this period. When the c.e.s. has a shorter life-time the excitation of the two impulses cannot summate.

For the synaptic transmission in ganglia it is possible for the effects of two sucessive impulses within a time of about 0,01 sec. in the same fiber to summate and result in a discharge of the ganglion cell.

In the central nervous system it is not probable that this kind of summation can occur. It has not been possible to demonstrate any phase of lowered threshold following subliminal synaptic stimulation what lasts longer than the refractory period of the fiber.

2e.  *The two impulses are conducted in different axons.*

As the central excitatory state has a spatial extension it is capable of summation with similar events set up in synapses in the immediate neighbourhood.

For the synaptic transmission in ganglia the impulses must arrive at the ganglion cells within about 2 msec, whereas the spatial extension in which the summation occur is rather large.

In the central nervous system it is not probable that this kind of sum- is possible does not exceed 0.5 msec.

The spatial extention in which the summation occur is located in discrete zones of the cell membrane. The number of impulses necessary for discharge varies but is at least two. Threshold excitation would occur when all, or at least a majcrity, of the knobs at a discrete zone are activated within a time interval of about 0,5 msec. Activation of sca'tered synapses would produce only a lowering of the threshold at corresponding loci cn the surface. If impulses from ano'her source activate the remaining synapses of these loci, discharge would result.

*Out of these data it is evident that the number of impulses required for vision must exceed one.* Indeed, we found that two impulses caused by the absorption of two quanta obeying the two quanta conditions succeed in arising a sensation.

*However, it is very remarkable that only two are sufficient.*

There are several successive neurons interlocated between the receptor neuron and the area of the brain in which the sensation is supposed to become conscious. In view of Sherrington's and Eccles' data, it seems to be that the excitation of two impulses cannot reach the third neuron. As indeed the third and other successive neurons are stimulated at the threshold of vision, perhaps impulses originating from processes of the two exciting impulses via the horizontal cells, according to the three principles mentioned in section 1., overcome the resistances of the second and other successive synapses.

In our opinion, the data available from nerve physiological investigations are not sufficient to draw detailed conclusions about the way in which the two impulses succeed in giving rise to a visual sensation.

*Anyhow the results of investigations from two fields of science, nerve physiology and physiological optics are similar: the visual sense organ and simple reflex arcs are excited by two impulses.*

# V. THE SENSITIVITY DEPENDENCE ON WAVE-LENGTH FOR FOVEAL AND PERIPHERAL VISION AND THE FUNDAMENTAL RESPONSE CURVES.

## 1. Introduction.

The initial process of the visual light-perception is the absorption of light-quanta by the photo-chemical material in the receptors of the eye. The sensitivity of the eye as a function of the wave-length is determined by the absorption power of the several photochemical substances present under the circumstances concerned in the receptors of the illuminated area of the retina and the specific losses of light by reflections and absorptions in the various parts of the eye. Moreover, there may be a dependence on the time of observation or on the visual angle of the test-spot.

From our threshold measurements for various wave-lengths described in chapter III [3]) it has appeared that for foveal as well as for peripheral vision the dependence of the threshold-values on the time of observation was the same for all wave-lengths. For that reason the sensitivity curve of the eye at the threshold of vision does not depend on the time of observation. It was also found that for foveal vision there was only a very slight dependence on the wave-length of the threshold curve as a function of the visual angle. The foveal sensitivity curves are practically the same for all visual angles of the testspot so long as these do not exceed the area of the fovea.

The peripheral threshold measurements proved the existence of a very strong dependence on wave-length of the threshold curve as a function of the visual angle, in accordance with the data of other investigators. [29]) This was explained in chapter III by the fact that in the fovea for the ligth-perception two quanta must be absorbed within a time of about 0,04 sec and within a distance corresponding to a visual angle of 2 to 4 minutes for all wave-lengths whereas, 7° nasal from the fovea, two quanta must be absorbed within ca 0,04 sec within a distance of about 12' for rod vision and within a distance of about 3' for cone vision. This means that the shape of the sensitivity curve at the threshold of vision is indeed dependent on the visual angle subtended by the testspot. In the region within which the sensitivity of the rod system is of the same order of magnitude as that of the cone system in the red part of the spectrum, the shape of the sensitivity curves for small and large visual angles will differ.

We extended our threshold measurements of chapter III for a few visual angles for peripheral vision as well as for foveal vision with the determination of the threshold-values for a large number of wave-lengths. For the experimental arrangement we refer to chapter III and likewise for the intensity-measurements. We define the threshold $\overline{N}_{60\%}$ as the average number of quanta of a flash, necessary for 60 % chance of observation. The measurements were performed by the normal trichromatical observer H.v.d.V. and by the deuteranomalous observer M.A.B.

## 2. The peripheral sensitivity measurements.

In fig. 17 we give the sensitivity defined by the reciprocal value of the threshold $\overline{N}_{60\%}$ obtained 7° nasal from the fovea for the dark-adapted right-eye and for the visual angles of 4' and 85' of the test spot for M.A.B. and of 4' for v.d.V. It is evident, that the two curves for M.A.B. agree except in the region from 5600 Å to 6400 Å in which region the cone system becomes important for the 4' curve. The shape of the 4'-curves of M.A.B. and v.d.V. did not differ essentially.

From chapter III we known that on the blue side of this region the absorption of two quanta in two rods within a distance of about 12' and within about 0,04 seconds is necessary for the light-perception, whereas in the red region the two quanta must be absorbed within about 3'. With increasing wave-length in the region 5600 Å to 6400 Å the sensitivity will therefore decrease faster for large visual angles than for visual angles smaller than 3'. In the red region of the spectrum the cones are more sensitive than the rods. At the threshold of vision for these wave-lengths the colour sensation connected with the light-perception is red whereas in the other part of the spectrum the specific colourless rod-sensation is seen.

The initial process of the light-perception for rod vision is the absorption of light-quanta in the rhodopsin of the rods. The absorption power of the rhodopsin in vitro has been measured by several authors. Their results agree fairly well. In fig. 17 we give the percentage absorption curve as determined by Lythgoe [30]). It refers to a dilute solution. Obviously the shape of his rhodopsin curve does not agree with our sensitivity curve for rod vision.

We must further combine the rhodopsin curve with the absorption in the various parts of the eye. For this purpose we can use the data of Ludvigh and Mc Carthy [31]) which are the best available. The resulting curve is also plotted in fig. 17. Especially in the extreme blue the absorption in the eye is rather large. A second reason for the deviation of the shape of the sensitivity curve from that of the percentage absorption curve of rhodopsin in a dilute solution may be that maximum absorption capacity of the

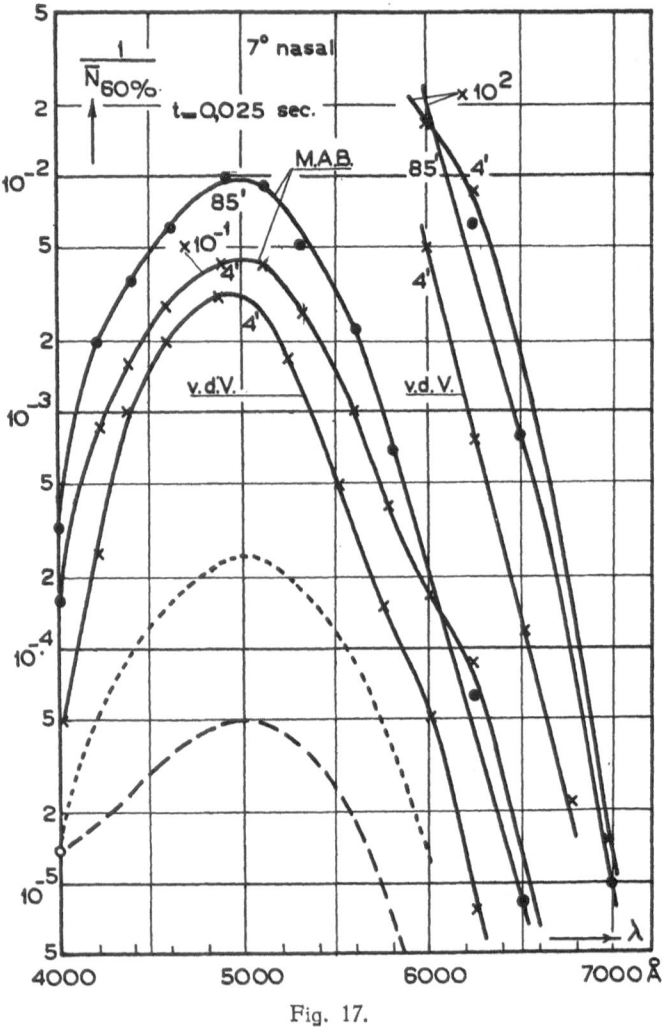

Fig. 17.

The peripheral sensitivity of the eye as a function of the wave-length for several visual angles; the shape of the percentage absorption curve of the rhodopsin (lowest) and of the percentage absorption curve of the rhodopsin combined with the absorption of the various parts of the eye.

rhodopsin in the eye is rather large. Indeed, the shape of the percentage absorption curve of a solution depends on its concentration. Hecht [16]) determined the effective absorption power of the rhodopsin in the eye by comparing the shape of the sensitivity curve, corrected with the data of Ludvigh and Mc Carthy, with the percentage absorption curves of the rhodopsin for several maximum absorption powers. In the region of small maximum absorptions this method is very insensitive. From the data of Ludvigh and Mc Carthy it appeared that about half the number of quanta of the wave-length of maximum absorption of the rhodopsin reaches the retina and Wald found [19]) that about 20 % of this fraction is effectively absorbed, so that about 10 % of the total number of quanta incident on the eye contributes to the light-perception.

Fig. 17 shows that this fraction is for M.A.B. 10 % and for v.d.V. 7 % as the optimal number of quanta are respectively about 20 and 30 and as two quanta are necessary for the light impression.

In fig. 17 on the blue side of the maximum the shape of the sensitivity curve agrees fairly well with the resulting curve for small maximum absorptions. On the red side of the maximum the slope is even steeper than agrees with the rhodopsin curve. The pigmentation of the retina can have influenced this region. Anyhow, it seemed in our opinion to be impossible to determine by this method the actual maximum absorption with great precision. The agreement of the sensitivity curve with the curve for small maximum absorption of rhodopsin is satisfactory for the greater part of the spectrum and we can only draw the conclusion, that the maximum absorption does not exceed about 30 %.

### 3. The foveal sensitivity-measurements.

In fig. 18 we give the foveal sensitivity-values obtained for the dark adapted right eye. As the distances within which the two quanta, necessary for the light-perception, must be absorbed do not differ essentially for the various wave-lengths (about 2 - 4 minutes) the foveal sensitivity curve is almost the same for visual angles below 48'. The minimum number of quanta at the threshold is foveal about ten times the corresponding number in the periphery 7° nasal from the fovea.

The shape of our foveal curves in which two peaks occur differ from the curves usually found from luminosity experiments. In the luminosity experiments of Wright (1946 page 79) also a rather small peak occurs at 6000 Å, but most of the investigators give curves without peaks. In our opinion a curve without peaks is more striking than with peaks as the sensitivity curve is the addition of several sensitivities due to separate systems, by which the facts of colour vision must be explained.

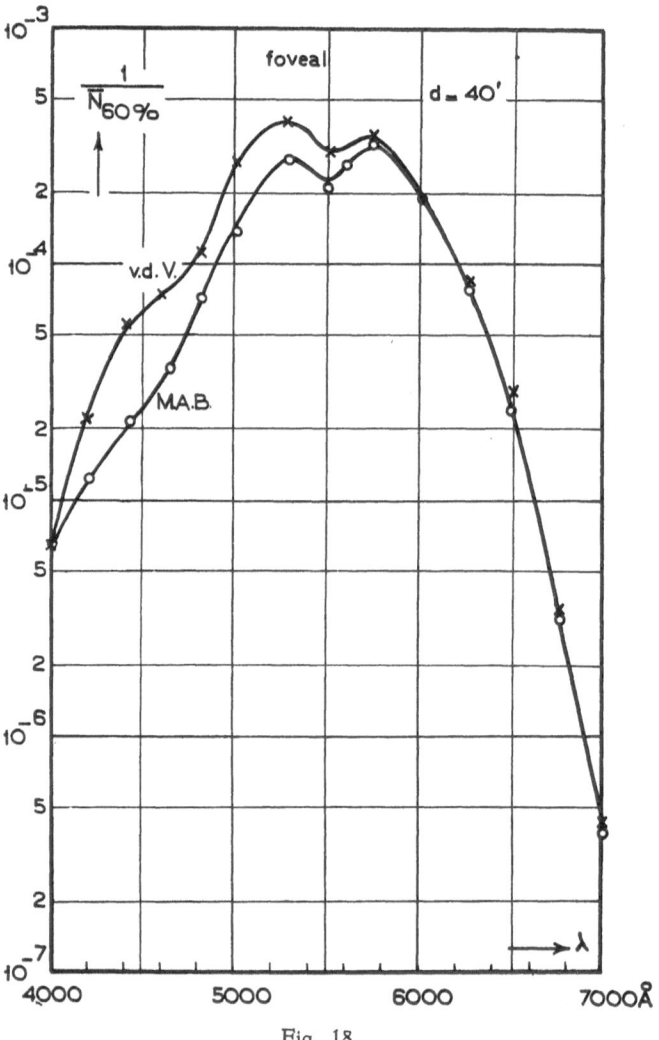

Fig. 18.

The foveal sensitivity of the eye as a function of the wave-length for d = 40' for the normal observer v.d.V. and the deuteranomalous observer M.A.B.

It seemed to us, that the determination of the sensitivity curve with the aid of the absolute threshold-measurements might introduce fewer complicating facts than more intricate investigations such as luminosity experiments with the aid of a flickerphotometer or simular arrangements.

The data of threshold-measurements published by Wald [32]) are averages of a great number of observers; this may have obliterated the peaks between 5000 and 6000 Å.

Thomson and Wright [29]) investigated the sensitivity curves for different points of the fovea with the aid of luminosity experiments. They found differences between the curve for the centre of the fovea and the curves for excentral positions of the test spot. The differences are maximal at 5500 Å and are in this region of wave-lengths between the central curve and the curve for an excentral position of 20' about 30 %. The largest visual angle of the test spot in our foveal measurements was 48', so that places about 25' excentral were just subtended.

The average sensitivity of a central area with 48' diameter can differ about 20 % from the sensitivity of a very small central area, according to Thomson and Wright's data. This difference depends on the wave-length. By this reason the shape of the threshold curve as a function of the visual angle of the test spot will be slightly different for the various wave-lengths. These differences were also noticed in our measurements (see page 47). The two quanta conclusion is not affected by this fact, as only a very small deviation from the theoretical two quanta curve is caused.

The shape of the sensitivity curve will be determined by the intrinsic sensitivities of the several kinds of cones and the specific absorption of the various parts of the eye.

For rod vision the photo-chemical material responsible for transmitting the energy of the quanta to nerve-impulses is known. It appeared that with the data of Ludvigh and Mc Carthy on the absorption of the various parts of the eye the peripheral curve can be made to agree fairly well with the absorption curve of the rhodopsin [30]).

Applying the correction for absorption of the eye (compare fig. 17) also to the foveal curve, we obtain the curve representing the sensitivity, as a function of the wave-length, of the quanta that reach the retina. Fig. 19 shows these curves. They differ rather much for the two observers. This may partly be due to differences in the absorbing properties of the various parts of the eye in front of the retina. Since, however, in the periphery at 7° nasal the shape of the curves agrees for the two observers, this is very improbable. Another reason may be the differences in pigmentation of the retina: a yellow pigment is concentrated in a zone about the fovea: the *macula lutea.* The absorption curve of this yellow pigment has been determined by Wald [32]) and agrees with the properties of xanthophyll. This macula pigment absorbs more particularly in the region 4900 - 4200 Å. In this region of wave-lengths the curves of v.d.V. and M.A.B. differ very much and this may be due to a larger concentration of pigment in the eye of M.A.B. than of v.d.V. Illuminating the retinae of the two observers with a Hg-light source in order to eliminate the red colour of the blood, the macula lutea of the eye of M.A.B., seen via the opthalmoscope *), has a more yellow colour than that of v.d.V. This agrees with the suggestion that the difference in shape of the curves of fig. 19 in

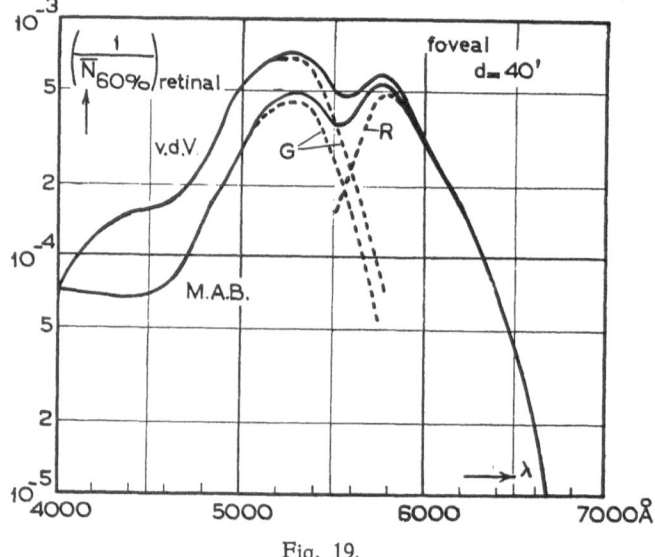

Fig. 19.

The foveal sensitivity of the retina as a function of the wave-length for the normal observer v.d.V. and the deuteanomalous observer M.A.B. and our "red" and "green" fundamental response curves.

the blue region may be due to different pigmentation. In the region of wave-lengths beyond 5000 Å the sensitivity curve is determined by the intrinsic sensitivities of the separate cone-systems, as for these wave-lengths no specific absorptions occur in the various parts of the eye.

It is not possible to decompose the curve into the components without speculations, but it is very probable that two of the cone-

---

*) Our thanks are due to J. ten Doesschate for these observations.

systems have a maximum absorption at about 5300 respectively 5800 Å for both observers.

It is likely from the curve for v. d. V. in fig. 18 that on the blue side a third cone-system have a maximum absorption in the region of 4400 Å, but in the region below 4800 Å the macular pigment will have changed the shape of the intrinsic sensitivity curves of the cones considerably. Wald [32]) found an average absorption percentage in the macula pigment of about 60 % dependent on the test person, but 90 % also occured. Combining these data with the curves of fig. 19, the intrinsic sensitivity of the cone systems proves to have a third maximum at about 4350 Å. For making the shape of the resulting curve in the region of 4200 - 4600 Å for v.d.V. and M.A.B. almost equal, the absorption of the pigment must be for M.A.B. at least 60 % and for v.d.V. 40 % in agreement with the higher pigmentation of M.A.B.'s eye. In Thomson's [33]) determination of the sensititity curve between 4800 and 6500 Å with the aid of the luminosity step- by step method resulting curves similar to ours are sometimes found.

In Thomson's data also a maximum at ca 6000 Å sometimes occur. Allthough a rather good qualitative agreement between his and our data exist, an extensive study on the disagreements seems to be necessary.

The ratio between the sensitivities at 5300 and 5800 Å is 1,25 for v.d.V. and 0,94 for M.A.B.

From the threshold-measurements it seemed possible that the anomaly of M.A.B. is partly due to a diminished sensitivity of the normal „green" receptor. *Anyhow no shift in the scale of wavelengths occurs of the peak at 5300 Å and 5800 Å for the deuteranomalous observer.*

*So it seems to be probable that the photochemical substances of the red and green receptors are for both test persons the same. But in this case the anomaly in the colour mixture data of an anomalous observer have no photochemical foundation. By this reason anomaly in trichromatic colour vision might be of nervous origin.*

Our results are in contradiction with the fundamental curves, found by de Vries [34]) with a method similar to that of Stiles [35]) experiments. De Vries postulated a general anomalous curve acting as red curve for the protanomalous and as green curve for the deuteranomalous observer. The maximum of this anomalous curve is situated at about 5600 Å whereas de Vries [34]) normal green and red curve have their maxima at about 5450 and 5650 Å.

Wright *) gave the reliable data available concerning the fundamental response curves. Among the results of the several more recent investigations the curves of Hecht [36]) are certainly not in agreement with our sensitivity-curves as their three maxima almost coincide.

The half-width value of the sensitivity of the "red" respectively the "green" curve is for Stiles [35]) 1100 and 1000 Å, for Wright's curves 1200 and 1000 Å, in Walters' [37]) work 1100 and 800 Å and from Pitt's [38]) data 1100 and 1000 Å, whereas in de Vries' [7]) curves these values are 1000 and 800 Å. In ours they are 500 and 600 Å.

The wave-length at which the sensitivity of the red and green receptors are equal for the several authors mentioned are respectively about 5300 Å, 5500 Å, 5700 Å, 5700 Å and 5500 Å.

As the shape of the sensitivity curve beyond 5000 Å is practically not influenced by specific absorptions in the media of the eye it is very probable from our sensitivity curves of figures 18 and 19, that for the normal observer v.d.V. this equal sensitivity will occur at 5600 Å.

From chapter III we know that the receptors of the human eye act in complete mutual dependence, in particular: one quantum absorbed in a receptor of a certain kind and one quantum in a receptor of another kind results in a light-perception. The resulting change of observation for two systems is according to the two-quanta theory and the mutual dependence:

$$W(\overline{N}) = 1 - (1 + f_r \, a_r \, \overline{N} + f_{gr} \, a_{gr} \, \overline{N}) . e^{- (f_r \, a_r + f_{gr} \, a_{gr}) \, \overline{N}}$$

$a_r$ and $a_{gr}$ are the fractions of the illuminated area of the retina in which red and green receptors are present. When the receptors are not homogeneously distributed $a_r$ and $a_{gr}$ depend on the visual angle. From chapter III is has appeared that when $d$ is at least 4 minutes, this dependence certainly does not exist, as the shape of the sensitivity curve does not depend on the visual angle.

Of course, $a_r$ and $a_{gr}$ do not depend on the wave-length. $f_r$ and $f_{gr}$ are the intrinsic sensitivities of the red and green cones and proportional to the sensitivity of the red and green receptor systems $f_r a_r$ and $f_{gr} a_{gr}$.

At the wave-length at which the two systems are equally sensitive, the resulting sensitivity has twice the value of the separate systems. When the ratio

---

*) Wright, 1946: page 368.

between the sensitivities of the two systems differs from unity, the sensitivity of the more sensitive system always exceeds half of the resulting sensitivity.

The fundamental response curves presented by the dotted line are highly conjectural, but there is no doubt that it is impossible to decompose the resulting curve into curves of which the half-width-values differ very much from the curves presented by us.

For the red this width is 500 Å and for the green 600 Å. The widths of our curves are very much less than those, given by the other authors mentioned. The study of the sensitivity curves for the different types of colourblind-persons, will give more information about these problems. We are studying the data for a tritanope and others.

The *modulators* found by Granit [39] from electro-physiological experiments have half-width-values of the same order of magnitude as ours. Granit suggests that the response-curves found by Pitt and others represent average-curves for the very large number of red and green receptors, which exist in a rather large area of the fovea. Considering our results, we are of the opinion, that the intrinsic sensitivity curves of the cones are probably as narrow as the modulators found by Granit. Though not likely, the possibility can not be excluded, that the red curve is itself a composite curve. Granit has built up his red response curve by combining two modulators, one with its maximum sensitivity at about 5800 Å and the other at 6000 Å. The slight bend at 6100 Å in the increase of the sensitivity from 7000 Å to 5700 Å in fig. 18 might be due to the actual existence of two modulators in the red region.

In the electro-physiological work mentioned two kinds of sensitivity curves were found, namely very wide ones, similar to the luminosity curve of the eye and rather narrow ones. Granit called the first: *dominator* and the second: *modulator*. In the greater part of the trials the micro-electrode found the dominator and in a comparatively small number a modulator. The modulators represent the intrinsic sensitivities of the individual receptors and must show the absorption-spectra of the photochemical materials in the cones. *In view of the two-quanta theory and the mutual dependence of the several kinds of receptors it is clear that all nerve elements of the retina have a wide sensitivity curve as the result of the dependence of the several kind of receptors connected with the element, except the nerve-fibers, which are directly connected with the cone- and rod-bodies.* Besides these nerve elements all others are stimulated when a quantum is absorbed in a receptor of the area of the retina with which they are connected, no matter what kind of cone or rod. All these elements will act in the way of a dominator. Only a very small part of the nerve-elements will act like a modulator, namely only then, when the micro-electrode is placed on the fibre of a rod- or cone.

In Granit's work and in our threshold-measurements the colour-sensation of the light-impression is not taken into account. The other investigations mentioned are studies on the colour-sensation and are thereby more complicated than Granit's and ours, which are of an all- or none-nature.

Granit determined in an electro-physiological way whether a certain receptor reacts or reacts not on a light stimulus of varying wave-length. All our experiments are based on observations whether a flash was seen or not, no matter the brightness, colour or hue impression.

For the interpretation of the experimental data obtained with more complicated methods basic assumptions on the mechanism of the visual sensation have to be made as for instance the hypothesis that in comparing brightness-impressions of different colour in the flicker photometer the sensitivity of the eye for a certain wave-length is proportional to the reciprocal of the intensity necessary for the disappearance of the flickering effect. This hypothesis treated on its merits objectively can be regarded as radical and probably subtends all aspects of vision.

In conclusion we remark, that the deviations from the "additional law" for small visual angles suggested by Bouma [40]) must be due to the chromatic aberration of the eye.

He found that the threshold energy for a mixture of light of different wave-lengths for small visual angles of the test spot is smaller than the value deduced from the thresholds of the separate kinds of light. Of course, the shape of the sensitivity-curve for peripheral vision as a function of the wave-length depends on the visual angle but when the right curve is chosen no devations from the additional law for mixed light occur provided the necessary precautions against the aberrations are duly considered. When these precautions are neglected deviations will naturally occur but these are of a secundary nature and do not constitute one of the fundamental properties among those of the processes of the retina. which lead up to consciousness, but they belong only to the focussing properties of the eye. *The completely mutual dependence found by us includes the impossibility of fundamental deviations from the additional law at the threshold of vision.*

# VI. THE QUANTA EXPLANATION AND THE BRIGHTNESS-IMPRESSION FOR VARIOUS TIMES OF OBSERVATION AND VISUAL ANGLES.

As it can be expected that the effects described in the two-quanta explanation are important for the brightness impression, we performed experiments suitable for the study of these phenomena concerning brightness impression.

We determine the dependence of the average number of quanta necessary for a certain brightness impression on the visual angle and time of observation for several values of the subjective brightness 7° nasal from the fovea.

## 1. The experiments.

For the experiments concerning the dependence on the time we asked the observer to compare the brightness impressions of two successive flashes of different duration but equal visual angle. The observer had to change the intensity of one of the flashes till he was unable to see any difference in brightness impression of the two flashes.

The tungsten ribbon filament lamps *I* and *II* were imaged by the lenses $a_1$ and $a_2$ on the screen of the disk *b*, in which two openings were made, so that the light of *I* could only pass through the one,

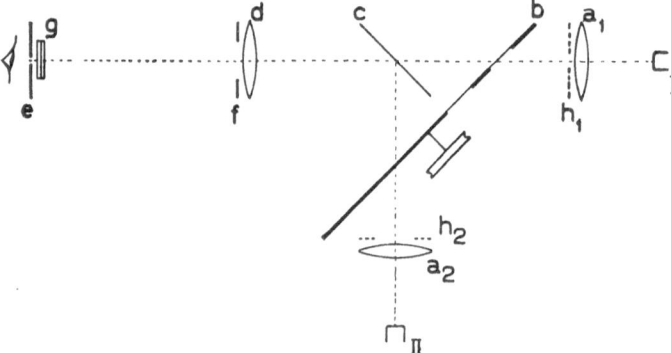

and that of *II* through the other opening. The images of *I* and *Ii* via the mirror *e* are thrown by the lens *d* on the artificial pupil *e*. The size of the openings in the disk and the velocity were such as to make the flash-time for the light of *II* always 0.1 sec. The flash-time for the light of *I* could be regulated between 0.05 sec.

and 1 sec. The time between the two flashes was always about 0.5 sec. The screen of the lens $d$, completely filled up by the light of $I$ or $II$, is the object for observation. The visual angle is determined by the diaphragm $f$. In front of the pupil $e$ a Schott $VG_2$ filter $g$ was placed with a maximum transmission at 5300Å. The head of the observer was fixed by means of a mouth-mounting piece. The dark-adapted right eye was used and the spot of the retina on which the flashes were received was situated 7° nasal from the fovea, just like in all our previous experiments. Fixation of the right eye was performed by the observation of a red fixation light with the left eye.

For several intensities of the test flash of 0.1 sec. of lamp $II$ the observer adjusted the currents of lamp $II$ to every fixated time of the flash of $II$ for which the brightness impression of the two flashes seemed to be equal. Every measurement was repeated several times and the average current value was determined. The experiments were performed for various visual angles by changing the diaphragm $f$.

In Fig. 20a and 20b we give, for some visual angles, the average number of quanta necessary for the same brightness impression as a function of the flash-time.

The same experimental arrangement was adapted to the experiments concerning the dependence on the visual angle.

The diaphragm $f$ was removed. The visual angles of the two flashes for which the time was now always equal, were determined by the two diaphragms $h_1$ and $h_2$ of the lenses $a_1$ and $a_2$. The beam due to flash $II$ has always the visual angle 30', as regulated by the diaphragm $h_2$, while the visual angle of the beam due to flash $I$ can be varied from 5' to 200' by means of the diaphragm $h_1$. The condition of the eye and its fixation were quite similar as in the experiments described concerning the dependence on time.

For several intensities of the test flash of 30' of lamp $II$ the observer adjusted the currents of lamp $I$ for which the two flashes seemed to be equally bright for every fixated visual angle of the flash of $I$. Again the average of several measurements was determined. The time between the two flashes was again about 0.5 sec. In Fig. 21 we give the curves indicating the dependence on visual angle of the average number of quanta of a flash for rod vision for a certain brightness impression.

Fig. 22 and 23 show the results obtained when the green filter

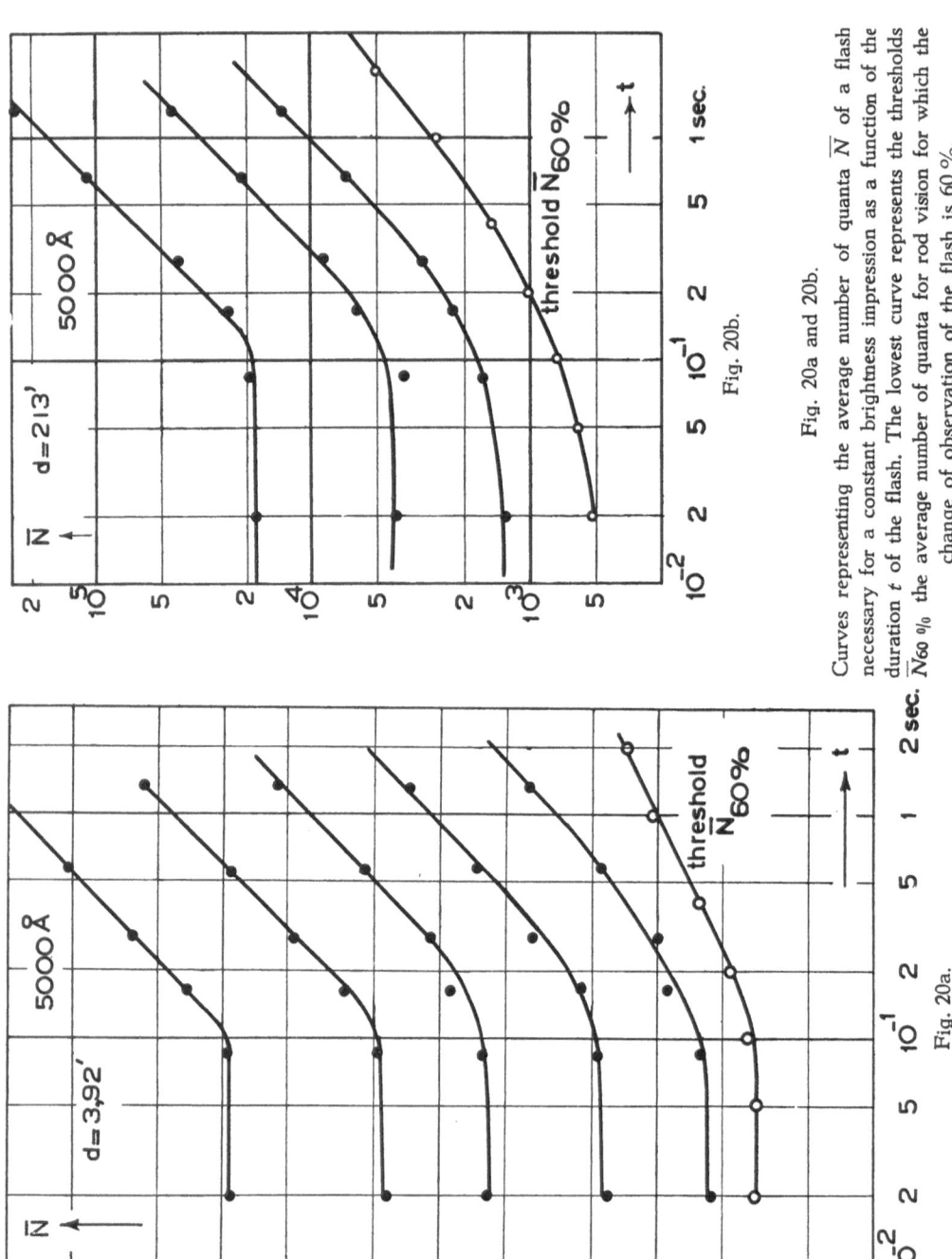

Fig. 20a and 20b.

Curves representing the average number of quanta $\overline{N}$ of a flash necessary for a constant brightness impression as a function of the duration $t$ of the flash. The lowest curve represents the thresholds $\overline{N}_{60\%}$, the average number of quanta for rod vision for which the change of observation of the flash is 60 %.

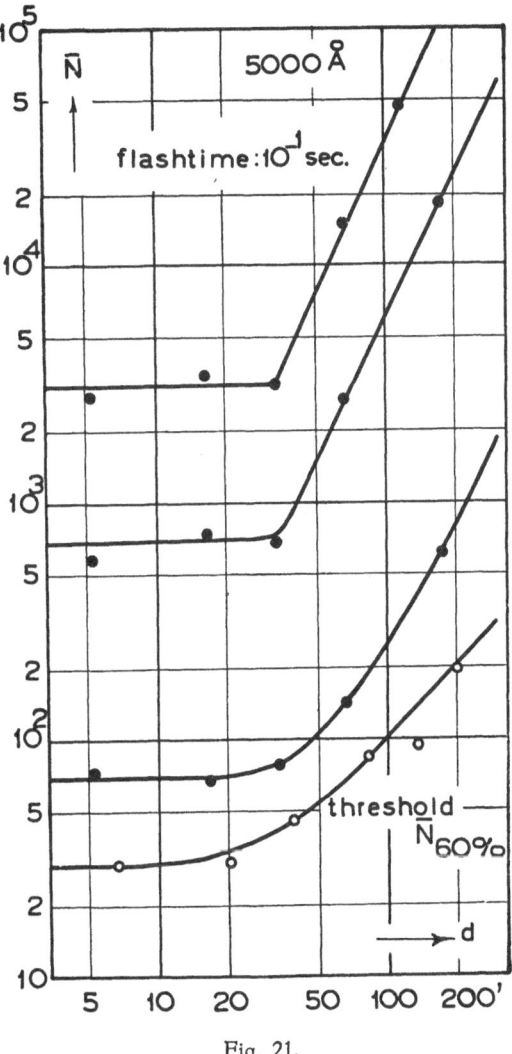

Fig. 21.

Curves representing the average number of quanta $\bar{N}$ of a flash necessary for a constant brightness impression as a function of the visual angle of the flash for rod vision.

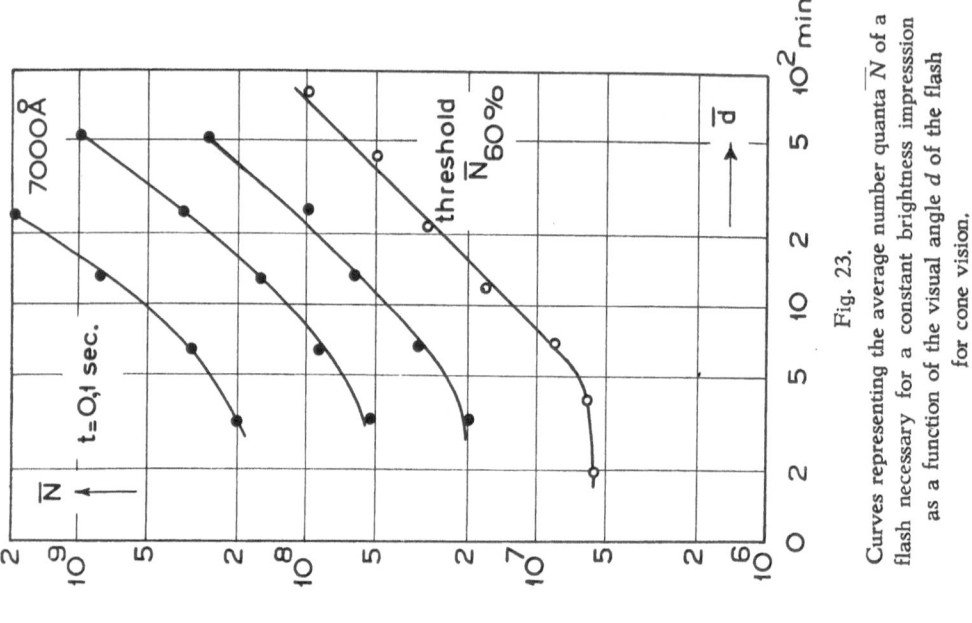

Fig. 23.

Curves representing the average number quanta $\overline{N}$ of a flash necessary for a constant brightness impresssion as a function of the visual angle $d$ of the flash for cone vision.

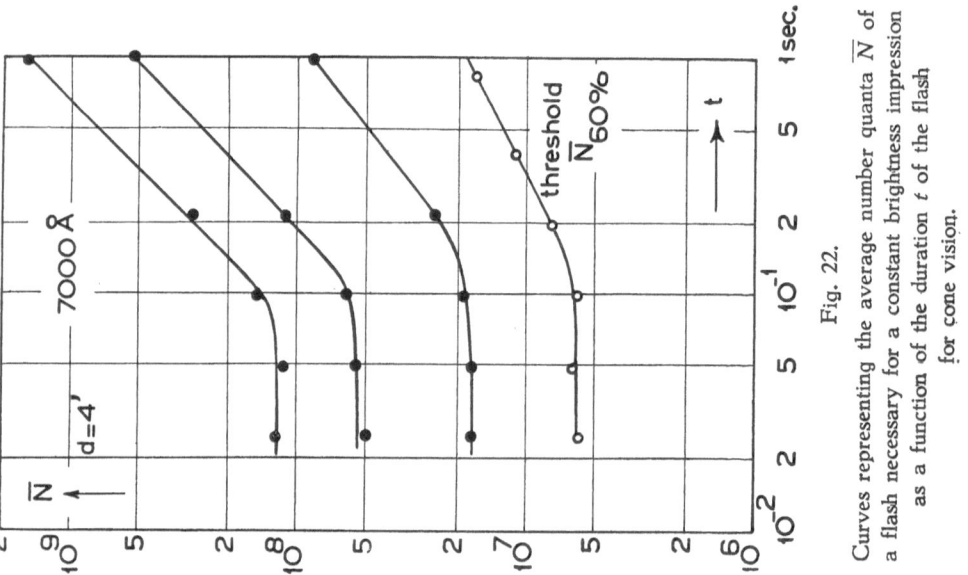

Fig. 22.

Curves representing the average number quanta $\overline{N}$ of a flash necessary for a constant brightness impression as a function of the duration $t$ of the flash for cone vision.

$g$ with which the rod vision data of figures 20 and 21 were found, was replaced by a red filter by which only wave-lengths beyond 6800 Å were transmitted, so that only the "red" cones were activated. See figs. 22, 23.

## 2. Discussion.

*α. The Dependence on Time.*

When in an area $D$ a very small number of quanta per second is absorbed, by stastistical fluctuations in the time between two succeeding absorptions, there will be a slight chance that more than one quantum is absorbed in a time $\tau$. When this happens light is perceived by the cooperation of the nerve impulses. The duration of this process of an elementary light impression begins at the moment the first quantum is absorbed and ends when the light impression disappears. The brightness-impression and duration will depend on:

*a.* The accidental number of absorbed quanta and the accidental distribution in time of the absorptions. We know that two absorptions within $\tau$ sec. cause a light impression, but about the influence of extra absorbed quanta on the brightness impression we know nothing.

*b.* The way in which the pulses of the activated rods are conducted by the retina, optical nerve, brain to consciousness. Even if the same receptors as regards time and place, are activated in exactly the same way, it is not sure that the way of conducting is also the same, so that light impressions are not necessarily identical in duration, extension, and brightness.

For very low intensities of the illuminated area of the retina the frequency of the elementary light impression is low. The average time between the beginning of the separate impressions is determined by the chance for the absorption of two quanta within $\tau$.

For increasing intensity this time decreases, and for a sufficient increase it will become smaller than the duration of an elementary light impression. For these intensity values we may expect singularities to occur in the behaviour of the brightness impression because the cooperation of the processes of elementary light impressions becomes possible. For the comparison of brightness impressions only the established equal or "just different" values are suitable. For the study of possible singularities one can investigate how a certain brightness impression is obtained for several times of observation.

It is not sure that we need the same intensity in order to obtain the same brightness impression for different times.

In Fig. 20a and 20b we show this dependence for two visual angles and several brightness impressions. Each curve represents the average number of quanta necessary for a fixed brightness impression. We give also the threshold curve $\overline{N}_{60\%}$

For increasing flash-times the number of quanta per sec. decreases for the threshold $\overline{N}_{60\%}$ The average time between the two absorptions which give rise to the light impression increases, and the chance for extra absorptions within a certain time of the absorption of the two quanta decreases.

In the lowest curve for a constant brightness impression of Fig. 20a the chance of observation of the flash is still $< 1$ for $t$ values $< 1$ sec. The frequency of the elementary light impressions is small and they will be seen separately. The moments for the perception of light are completely determined by the chance for the absorption of two quanta within $\tau$ sec. The shape of the curve deviates from the curve $\overline{N}_{60\%}$ when $t > 0.1$ sec. We can conclude that for a certain brightness impression of the visual process of an elementary light impression the average number of quanta per sec. may not decrease below a definite value, as the deviations are such that for values of $t$, for which $\overline{N}_{60\%}$ is still proportional to $t^{1/2}$, the number of quanta for the constant brightness impression becomes proportional to $t$.

It can be concluded from the curves for higher values of the brightness impression where continuous light is perceived, that *for a certain brightness impression the average number of absorbed quanta within 0.11 sec. must be constant; as when the flashtime is smaller than 0,11 seconds the average number of quanta $\overline{N}$ is independent of the time and when $t > 0,11$ seconds $\sim t$.*

From fig. 22 it is seen that the behaviour for cone-vision is quite similar for the dependence on time.

### β. The Dependence on the Visual Angle.

For the elementary light impression discussed under α, the points *a* and *b* will also be important for the extension of the impression. When a flash of very low intensity is given having duration $t \leq \tau$, and visual angle which is large compared with $D$, there will be a chance for the absorption of more than one quantum by the statistical fluctuations in an area $D$. When this occurs, the nerve-impulses will produce an elementary light impression. The average distance of these impressions is determined by the chance just mentioned. With increasing intensity this chance increases, so that the average distance between the elementary light impressions decreases. An increasing part of the illuminated area will seem to be filled with light. The possibility of cooperation of the elementary impressions increases. It is now important to know the average number of quanta as a function of the visual angle for a constant brightness impression.

The data concerning this dependence are given in Fig. 21 for rod vision. In a way quite analogous to the time-dependence, we can conclude that the average number of quanta per $cm^2$ may not be smaller than a definite value, so that the effective absorption of more quanta, within a certain area of the place of absorption of the two quanta responsible for the occurence of the light impression, is important for the brightness impression. *It can further be concluded that for a constant brightness impression the average number of absorbed quanta within an area O with a visual angle o of about 25' must be constant as when o> 25' the number of quanta N is independent of the visual angle and when o> 25', $\overline{N} \sim o^2$.*

From fig. 23 it can be concluded that for "red" cone-vision o and so also O is smaller and o does not exceed 4 minutes of arc.

### 3. Conclusion.

Summarizing the results of the previous section: for a certain brightness- impression the average number of effective absorptions within 0.11 seconds within an area with 25' diameter for rod vision and 4' for cone vision must be constant. How must this property be explained?

The behaviour of threshold values $\overline{N}_{60\%}$ for small visual angles and small flash-times were described with the aid of the statistical fluctiations in time and place of the absorptions of the quanta in the separate rods and "red"cones: the two-quanta explanation. The nature of the chemical reactions of the photochemical material turned out to be unimportant for all employed intensities. (See page 35). For the measurements of figs. 20-23, especially for those curves for which the condition that the chance of observation exceeds 100 % is sligthly more than satisfied, the intensities are equal or only a little higher than those occurring in the threshold measurements of the previous chapters, *so that the behaviour of the brightness impression can also not be explained in a photochemical way.*

In our opinion, the hypothesis that the laws of Talbot and Ricco, being of a physiological nature, must be held valid. This hypothesis predicts that the stimuli within a certain area and within a certain time can cooperate. Within this region of $d$ and $t$ is $\overline{N}$ constant.

From the conditions for the occurrence of a light impression, it appeared that impulses cooperate within an area with diameter $D$ and time $\tau$. From the experiments on the brightness impression we can conclude that the number of impulses within an area $O$ (with a corresponding diameter $o$ of 25' for the rods and 3' for the cones) and a time $T$, caused by the same number of quanta, cooperate to create the brightness impression. As the first absorbed quantum does not give rise to a nerve impulse from which a light-impression results, one may doubt, the two quanta condition once being satisfied, whether *every* absorbed extra quantum causes a nerve impulse which increases the brightness impression.

From the two-quanta explanation for rod vision we know that the situation that arises after the effective absorption of one quantum in one receptor extends over a certain area of the retina during a time $\tau$. By this absorption this area is prepared for the occurrence of a light perception when a second quantum is absorbed in this area in a receptor, no matter which. So it proved that after the first absorption the impulses of a second receptor have the same result in view of the possibility of a light perception. This agrees with the data of Polyak [7]) that all recipient elements can influence each other by the nerve system.

However it can not be ascertained from the experiment whether the conditions for the development of the light perception once being satisfied by the absorption of two quanta all extra impulses in an area $O$ round the first two nerve impulses and within a time $T$ will be of the same kind in the quality of the light perception as regards brightness and colour. The brightness impression will generally increase when the average number of quanta increases, but it might be that for the increase of the brightness impression a mechanism exist, similar to the two quanta effect for the development of the visual sensation.

For "red" cone vision in this region of the retina the first two quanta must be absorbed in the selfsame receptor because of two possibilities: either that within the distance $D$ along which the impuls of the first quantum is transmitted by the nerve connections of the stimulated cone, no nerve fibers of a second receptor are situated or that the properties of the nerve connections are such that for "red" cone vision two quanta must always be absorbed in one receptor. Be it as it may from fig. 20 it proved that for brightness impression the area $O$ agrees almost with the area within which the two quanta condition must be satisfied so that all quanta necessary for a certain "red" brightness-impression in the extreme red must be absorbed within one cone within a time T. Especially in the region within which the rods and cones are equally sensitive the study of the colour sensation connected with the light impression is very important in relation to the problems of the "mixed-stimulus" and of contrast-sensitivity.

The behaviour of the threshold values did not enable us to decide for rod vision whether the two quanta must be absorbed within an area with diameter $D$ of the retina or within a dis'ance $D$ of each other.

In view of the study of Polyak the latter is the more probable one.

Now we cannot ascertain from the experiments of the brightness impression for rod vision whether the distance $o$ is the maximum distance within which impulses cooperate, or whether there is an area of diameter $o$ in the retina within which cooperation is possible. Here again the former is the more probable.

By the slight difference in the size of $D$ and $o$ it is, moreover, difficult to insure that the cooperation of all stimuli is already complete when 2 quanta are absorbed within a distance $D$, situated within $O$.

As for larger visual angles the two-quanta shape in the curves of the brightness impression disappears for lower intensities compared with smaller visual angles it seemed probable that one couple of quanta within $D$ is sufficient for the complete cooperation of all impulses within the surrounding $O$.

For foveal vision we have as yet not any experiments similar to those described in this chapter. Schouten [23]) found a behaviour of the brightness impression in the fovea which behaviour was explained by the $\alpha$-adaption. It can be expected that conditions are here more complicated than at the periphery.

The area $O$ has the nature of a recipient unit as the observer has always the same brightness impression when the total average number of quanta in this area is the same, regardless of the distribution of the places of absorption. Moreover, it is impossible to recognize a figure of which the details are smaller than the area $O$ of rod vision.

The size of the recipient unit found by ten Doesschate [42]) on this spot on the retina is smaller. The visual acuity, of which the upper limit is determined by the area $O$, was the basis for his computation. He used the experiments of Wertheim who did not apply the same test figure. Moreover, the area may differ for different persons.

The area $O$ was till now assumed to be well defined. As the curves of Fig. 20 show rather sharp bends, it is probable that the border of the area is rather well defined.

The time $T$ has the nature of an averaging time; the observer gets always the same brightness impression when the total average number of quanta within this time is the same, regardless of the distribution in time of the separate absorptions.

In our opinion the critical frequency, which is for the dark-adapted eye and high intensities in the periphery almost independent of the intensity and is about 10, is determined by the time $T$.

The time $T$, within which stimuli of the rods can cooperate, may not be sharply defined. As the curves of Fig. 21 are rather sharp-bended it seemed again to be probable that the time $T$ is also rather well defined.

# VII. VISUAL ACUITY.

One method, very often used for determinations of the visual acuity, consists of the recognition of two separate black stripes as given in Fig. 24. It is clear that visual acuity is closely related to

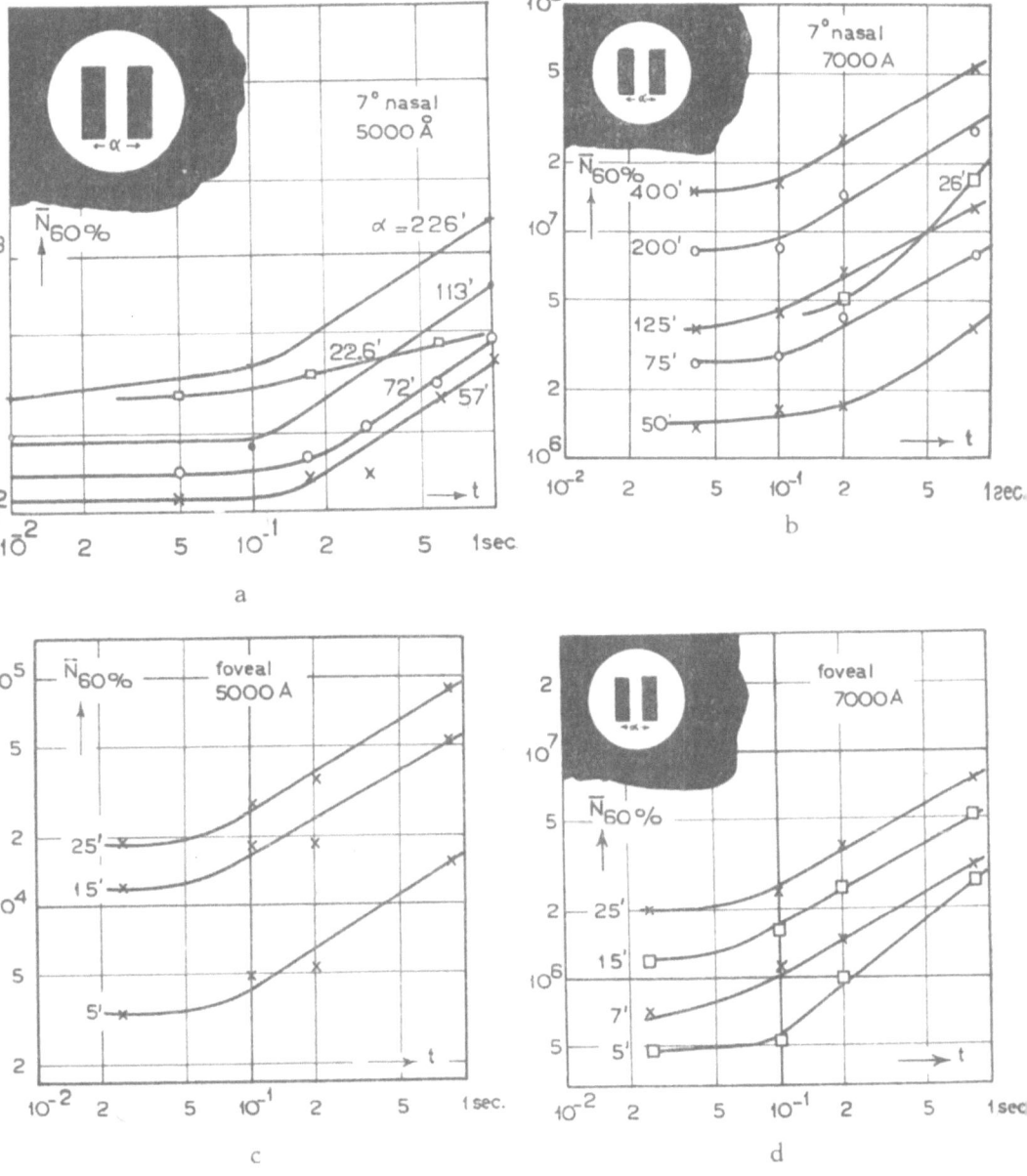

Fig. 24.

The number of quanta from the field between the two black stripes as a function of $t$, the time of observation, for different values of the visual angle $\alpha$ for a chance of recognition of 60 % for some wave-lengths for foveal and peripheral vision.

threshold values. The two stripes will be recognized as separately when the field between the two black stripes is seen. Obviously we can define only a certain chance of recognition just as in the case of threshold values. Moreover, it is evident that this chance of separate recognition will show a similar dependence on the visual angle and the time of observation as was established for the measurements of the threshold of perception. Again we take the chance of 60 % as a criterion.

*As soon as the visual angle of the details is of the same order of magnitude as the diameter of the sensitive unit, deviations will occur from threshold curves of the types shown in Figs. 11a and 11b.*

These deviations are due to the fact that within a sensitive unit separate ligth sensations cannot be observed at a low brightness. With the same arrangement as was used for threshold measurements, (in fig. 9 the shape of the diaphragm b was made to conform with the figure in fig. 24) we determined the relation between the chance of recognition, and the visual angle of the stripes and the duration of the flash. These measurements were performed for different values of the brightness and for several wave-lengths for foveal and peripheral vision.

The number of quanta from the field between the two black stripes is given in fig. 24 a-d as a function of the time of observation for different values of the visual angle covered by the field between the stripes. Comparisons were made for a chance of recognition of 60 %.

The curves for 22.6' for 5000 Å and 26' for 7000 Å for peripheral vision shows the existence of the area O.

It is noticed that the area O for "red" cone vision in the periphery is much smaller compared with the visual angle for which in fig. 24b deviations occur from the threshold curves of fig. 11 b. This is of course due to the fact that in an area of about 20' only one "red" cone is situated in this region of the retina. As soon as the visual angle of the details of the figure is of the same order of magnitude as the distance of the receptors deviations from the data of fig. 11 b will occur. This is demonstrated in fig. 24 b by the curve for 26'. Obviously the area O for the fovea is much smaller. It was not possible with our experimental arrangement to reach intensities large enough to obtain the data for 3', but there is no doubt that for this visual angle deviations from the ordinary threshold values from fig. 13 b occur similar to the deviations mentioned for peripheral vision.

86

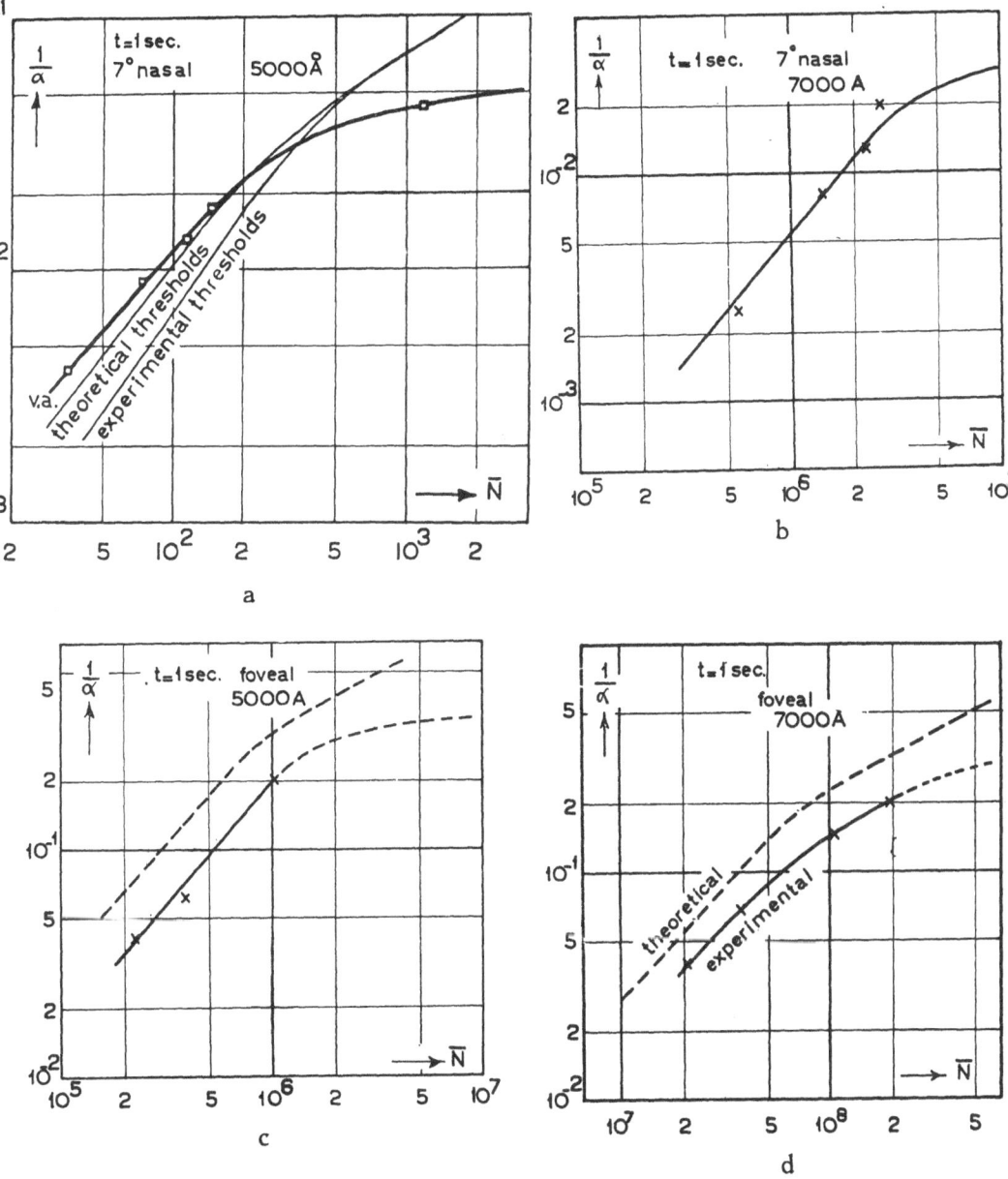

Fig. 25.

The reciprocal of the visual angle $\alpha$ (visual acuity) as a function of $\overline{N}$, the intensity, expressed in the number of quanta per second and per square area (laying on the retina) with a visual angle of $10^{-2}$ radian for some wave-lengths for peripheral and foveal vision. Some theoretical curves are given, using the threshold values from the two quanta theory and the threshold values found experimentally and plotted on an arbitrary intensity level.

From these experimental curves, one can derive the relation between the reciprocal of the visual angle (the visual acuity) and the brightness for a fixed duration of the flash, for example 1 second, as is shown in Fig. 25.

The analogous function was derived from the threshold measurements and from the theoretical threshold values from Fig. 8a and 8b, and is also given in Fig. 25.

Allthough we only performed extensive experiments on threshold values with the simultaneous occurrence of long and large flashes for 5000 Å, 7° nasal from the fovea, preliminary experiments showed that the deviations observed here are of a general nature and also occur for other wave-lengths and places of the retina.

From Fig. 25 we can conclude that the visual acuity is in better accordance with the thresholds derived from the two-quanta theory than with the thresholds found experimentally.

These facts are discussed at the end of the last chapter.

From the two-quanta theory it is clear that the total number of quanta $\overline{N}_{60\,\%}$ of a flash is proportional to the visual angle. At the threshold of vision the intensity $\overline{N}$ is therefore proportional to the reciprocal value of $x$ when $x$ is large compared with the diameter $o$ of the area $O$. The results of figures 25 a-d are in accordance with the theory.

De Vries[12]) has given a theory of the visual acuity, using the one-quantum hypothesis. His relation between visual acuity and intensity is not in accordance with the experimental curve. According to the one-quantum theory, the threshold values are not dependent on the visual angle. It would be expected in that case that the visual acuity is proportional to the square root of the brigthness which is contradictory to the experimental results.

Pirenne [43]) performed also some measurements on the visual acuity and determined what the intensity of a very large screen must be such that a round black spot in the middle of it is seen in 50 % of the cases.

So long as the visual angle of the black spot is large compared with the size of the recipient unit, we must expect that it is "seen" when parts of the surrounding area are seen. It would be stretching the point too much to deduce from the slope of the resulting curve for large visual angle the number $k$ required for the light-perception. In the experiments of Pirenne the centre of the screen was seen 20° out. The visual angle of the dark spot in the region of large angles ranges from 30° to 3°. It is quite sure that the sensitivity of the retina differs very much over such large distances. Moreover, in our opinion, realizing the presence of a black spot in such large surroundings is not very suitable for investigating the number of quanta required for the light-perception. For that purpose one must chose the most simple arrangement. Concerning the break in the curve of the visual acuity we did not find this in our visual acuity measurements, but especially in the region of small visual angles Pirennes data are more complete. In some of our preliminary experiments on brightness discrimination considered as a function of intensity we found a break. It might be possible that the cause for this break and for the sudden bend in the curve of Pirenne are the same.

Finally, we may still remark that the experimental results of Graham and Cook [44]) about visual acuity, and of Graham and Margaria [45]) and of Wald [46]) about visual threshold are in agreement with the results here presented in so far as their measurements, which neglected the consequences arising from the quantum nature of light, are comparable with ours.

# VIII.  ASPECTS OF VISION.

In the threshold measurements for the simultaneous occurence of large visual angles and large flash times, deviations from the predictions of the two-quanta explanation occured. These deviations were caused by physiological or psychological changes by the light of the test-flashes. In this chapter we discuss an aspect of vision and the other senses important for quantitative investigations.

For the study of sensations only the criterion of perceptible different impression whether in hue, brightness, colour, weight, heat, smell, loudness, etc. can be used. The difference of the impression can be examined by simultaneous or successive comparison. The one refers to the presence of different impressions at the same time the other to the change of the impression with time.

Here only brightness impression of the visual sense is discussed. We presume the observer is shown a small area from which the intensity can be increased from $\bar{n}_1$ to $\bar{n}_2$ in such a way that the time for which the intensity is increasing can be varied. According to the experiments about the brightness impression it proved that this impression is determined by the total number of quanta absorbed in a time $T$. When the time for which the intensity $\bar{n}_1$ increases to $\bar{n}_2$ is large compared with $T$ the brightness-impression will slowly increase from the one belonging to $\bar{n}_1$ to the one of $\bar{n}_2$. When this time is chosen smaller or equal to $T$ the brightness-impression will abrupt increase from the one of $\bar{n}_1$ to the one of $\bar{n}_2$.

The moment that the intensity $\bar{n}_2$ is reached the brightness-impression belonging to $\bar{n}_1$ is not present in the visual area. According to this a direct comparison of the two impressions is not possible. With a slow increase the memory of the first impression is obliterated by all impressions between $\bar{n}_1$ and $\bar{n}_2$. For the determination of the optimal threshold of seeing contrast between the intensities $\bar{n}_1$ and $\bar{n}_2$ the increasing of $\bar{n}_1$ to $\bar{n}_2$ must be performed within a time $T$. It is now also clear that when $\bar{n}_1$ is abruptly increased to $\bar{n}_2$ the two times $T$ just before and just after the increase are of special importance for the seeing of contrast. When a great part of the visual field has the intensity $\bar{n}_1$ and only in a part of this field is the intensity increased it is possible at every moment to compare the changing brightness impression with the original which is present all the time in the surrounding. The time for which $\bar{n}_1$ is increased to $\bar{n}_2$ in this case might be less important. It might be possible that for the seeing of contrast the change with time of the brightness impression on a definite place is more sensitive compared with the simultaneous comparison of brightness impressions on different places. Anyhow optimal conditions for the seeing of contrast are present when the change of $\bar{n}_1$ to $\bar{n}_2$, occurs in a time not greater than $T$, also when a surround is used.

Analogous to these problems of changes with time we can discuss the changes of the intensity with place in the visual field. When we present a flash $(t > T)$ of a test spot not homogenously illuminated with a large visual angle and in which $\bar{n}_1$ and $\bar{n}_2$ are the lowest and highest intensities the conditions of seeing contrast are optimal when the increasing from $\bar{n}_1$ to $\bar{n}_2$ is realised within a distance corresponding to the diameter $O$ of a recipient unit. When the intensities $\bar{n}_1$ and $\bar{n}_2$ are adjacent for the seeing of contrast the areas $O$ situated at the border are of special importance. Much prism's used in subjective photometry illustrate these fact by making the border as long as possible (Lummen Brodhun).

We may note that when before and after the flash the intensity $\bar{n}_1$ is present in the whole visual field the manner of changing in the field from $\bar{n}_1$ to $\bar{n}_2$ in the flash is less important, but optimal conditions for the seeing of contrast are present when the change from $\bar{n}_1$ to $\bar{n}_2$ is realised within a distance corresponding to the area $O$.

From the peripheral threshold-measurements for rod vision it is proved that the chance for a light impression is decreased by the absorption of quanta after $T$ seconds from the beginning of the flash over an area $O$. In chapter $II$ it was suggested that the reason for this impeded chance was either physiological or psychological. Actually we found that when to a subject are presented flashes with $t < \tau$ and $d < D$ on a background of which the intensity is just below the threshold value the threshold of the flashes is not increased compared with the threshold without a background. When the impeded chance was caused by the nerve-change of the retina by separately absorbed quanta, the threshold had to be slightly increased for a just not visible background (for in the visual field of the background a large number of quanta are absorbed). The increase need not to be large but would have to be noticeable.

In our opinion it is probable that the impeded chance for a light perception is caused by the absorption of couples of quanta satisfying to the two-quanta conditions.

Anyhow, at the threshold of vision we have the best possible circumstances for obtaining optimal sensitivity limited by the quantum nature of light and the physiological properties when $t < T$ and $d < o$, the diameter of the recipient unit. It seems that for obtaining threshold-values for $t > T$ and $d > o$ the beginning and end of the flash and the border of the test spot are of special importance, as they also are for the comparison of brightness impressions.

From figures 25 a-d it is seen that the deviations found in the threshold values are not very important for the measurements of the visual acuity with the test-figure of fig. 24. For large visual angles and flash-times the visual acuity is proportional with $\overline{N}$ in agreement with the two-quanta theory. We replaced the test figure of fig. 24 by two illuminated rectangular strips and repeated the visual acuity measurements. Deviations now occur similar to those found for threshold values. The difference in the behaviour of the visual acuity for these two test figures would seem to be due to the presence of an illuminated surrounding on the strips in the test figure of fig. 24. In this figure the brightness impression of the illuminated strips is stimulated by the absorbed quanta in the surround, according the effects discussed in chapter VI. These effects counterbalance the influence of the impeded chance for a light-perception discussed in chapter II. So it again proved that the border of the test spot can be of special importance for the visual sensation: it is efficient to make the border as long as possible.

The origin of these effects can be physiological or psychological. Some facts suggests that the deviations in the threshold-values when $t > T$ and $d > o$ are psychological in nature: it is rather difficult to give ones attention to the whole visual angle during the long time of the flash, so that light impressions can escape from full knowledge. In consequence the brightness impression of the visual field can increase unperceived so that for the observation of threshold values of these flashes the first and last time $T$ and the recipient units near the margin of the test spot become more efficient. In view of this the retina-grey might be of some importance. When the test spot is not circular the threshold value can become dependent on the ratio between outline and area as with increasing of this ratio the relative number of recipient units near the border will increase and by this the efficiency of seeing contrast [47]. It is evident that this possible explanation also holds for measurements of intensity, colour- and hue-discrimination. Similar aspects are of importance for the other human senses, especially on the time effect: in comparing the loudness of signals, the intensities of sources of heat, a.s.o. the observer always tries to present the different sensations many times in a quick succession in order to observe whether a change with time occur.

It is a fundamental problem whether the several facts discussed in this chapter are psychological aspects of the sense organ or a physiological aspect of the nerve-system or a mixture of both.

Several of these facts can be discussed in a psychological way but it is not impossible that these "psychological" considerations are a paraphrase of the transmission of impulses by the nerve-system.

The facts mentioned can generally be resumed by the conclusion that for the visual sensation the first and last part of the observation and the border of the observed test-spot are more efficient.

In view of the problems mentioned in this chapter electro-physiological experiments on optical nerve elements are of great importance.

Several authors [48], [39]) on electrophysiological investigations of nerve fibers reported observations of the activity of optical nerve fibers and found effects similar to the facts mentioned, namely the "on-effect", "off-effect" and "on-off effect". The first indicates a high activity at the beginning of the stimulation, the second at the end and the third at the beginning and the end.

It might be that this aspect of nerve-transmission has a close bearing on the problems of vision discussed in this chapter.

# IX. SUMMARY

In determining the smallest amount of energy necessary for vision fluctuating results are found: a flash originating from a constant light source is sometimes seen and sometimes not. Out of the work of van der Velden it proved that the fluctuating phenomenon mentioned is due to the statistical behaviour of the light quanta: a pure quantum physical aspect. From his investigation on the seeing of flashes he concluded that 7° nasal from the fovea for rod vision of the dark adapted eye independent of the observer two quanta must be absorbed in the visual purple within about 0,02 seconds and within an area of the retina corresponding to a visual angle of about 10' in order to perceive light.

As it appeared to be possible to distinguish between the physical and physiological aspect of vision, it can be expected that by a consistent extension of such experiments other fundamental properties of the behaviour of the visual sense organ can be found. To this purpose are the experiments described in this paper performed. In Chapter I the two quanta theory, its fundamental experiments and our experimental confirmation for rod vision are discussed. Chapter II gives an exposition of threshold experiments for rod vision in which a larger number of combinations of visual angle $d$ and flash time $t$ are studied. It proved that deviations from the two quanta theory occur when $t$ and $d$ are large.

These deviations can be explained by the fact that some time after the beginning of the flash the chance for a light perception is impeded. This state has developed after a time $T$ (ca 0,1 sec.) and extends over an area $O$ (about 20').

Chapter III deals with the investigation of cone vision. With the aid of the results of experiments on flashes on various places of the retina with monochromatic and heterochromatic light it is concluded that every receptor of the eye reacts at the absorption of a quantum with a nerve impulse to its nerve connection and a light impression is caused when a second quantum is absorbed in a receptor within a certain distance $D$ of the first receptor and within a time $\tau$ after the absorption of the first quantum.

The time $\tau$ is in our experiments independent of the place of the retina and of the kinds of excited receptors (0,04 seconds), so that it is probable that the interaction between the two impulses in the nerve connections of the receptors allways occur in a nerve element of the same kind.

The distance $D$ is in the fovea almost independent of the wavelength, so that this distance is almost equal for the several cone systems (2-4 minutes). 7° nasal from the fovea, $D$ is for the cone system sensitiv in the extreme red about 4' and for the rod system 10'. For the development of the light impression it is indifferent whether the two quanta are absorbed in receptors of the same kind or not, sothat a completely mutual dependence between the rod and cone systems in the periphery and between the cone systems in the fovea exist.

In Chapter IV some general proporties of the nerve system are discussed. The general principles of the histology of the nervous system and the nervous transmission are reviewed. In according with the mutual dependence found by us between the several cone systems in the fovea we concluded that there exist mutual connections between the cones, somewhere in the visual pathways.

Out of the work of Sherrington on the excitation of simple reflex arcs and of Eccles on the potential disturbances of nerve-elements it proved that the results of investigations from two fields of science, nerve-physiology and physiological optics are similar: the visual sense organ and simple reflex arcs are excited by two impulses.

In Chapter V experiments on the spectral sensitivity curve for rod and cone vision are presented. The shape of our foveal curves obtained with the aid of absolute threshold measurements differ from the curves usually found with luminosity experiments. In ours a few rather small peaks occur due to the intrinsic sensitivity curves of the seperate receptor systems. It seems out of the shape of the peaks that the half-width values of these intrinsic sensitivity curves are rather smaller and agree with these values for the "modulators" of Granit, contrary to the fundamental response curves found by other authors with luminosity experiments, which are all wider.

No shift in the scale of wave-lengths of the peaks for a deuteranomalous observer occurs, so that anomaly in trichomatic vision might be of nervous origin and has no photochemical foundation.

The deviations from the "additional law" for small visual angles suggested by Bouma must be due to the chomatic aberration of the eye. Fundamental deviations from the additional law are impossible at the threshold of vision.

In Chapter VI the behaviour of the brightness impression is studied with the aid of experiments on the comparison of the impression of flashes of different time and size. For peripheral vision it is found that for a certain brightness the average number of effective absorptions within 0,11 seconds within an area with 25′ diameter for rod vision and 4′ for cone vision must be constant. It is concluded that the impulses within the area and time mentioned cooperate to create the brigthness impression. The brightness impression will generally increase when the average number of absorptions increases. For "red" cone vision in the area O 7° nasal from the fovea only one "red" cone is situated, so that in the extreme red all quanta necessary for a certain brightness impression must be absorbed within one cone. For rod vision the area O contains about 400 rods. The area O has the nature of a recipient unit, the time $T$ of an averaging time.

Chapter VII deals with the visual acuity. As far as the visual angle of the details of the testfigure are large compared with the diameter of the recipient unit the behaviour of the visual acuity is similar to the behaviour of the threshold values and agrees with the two quanta explanation: at the threshold of recognition the intensity is proportional to visual acuity.

When the details of the test figure are of the same order of magnitude as the diameter of the recipient unit deviations from the two quanta behaviour for simple threshold experiments occur due to the fact that within such an unit separate light sensations cannot be observed.

In Chapter VIII some general aspects of vision are discussed in relation to the simultaneous and successive comparison of sensations. It is deduced that for the perception of differences in sensations the first and last averaging time $T$ of the observation and moreover for the visual sense the border of the test spot are of special importance. The consequences of these facts are confronted with the deviations mentioned in chapter II, problems of the visual acuity and the influence of the retina grey.

It is possible that the "psychological" considerations of this chapter are a paraphrase of the transmission of impulses by the nerve system and it is suggested that the on-effect, off-effect and on-off effect of nerve transmission has a close bearing on the problems of vision of this chapter.

*Physical Institute of the State University of Utrecht,*
*The Netherlands.*

# ZUSAMMENFASSUNG

Bei der Messung der zu einer Lichtempfindung nötigen Minimal-
energie werden wechselnde Resultate erhalten: Ein schwacher
Lichtblitz, erzeugt von einer konstanten Lichtquelle, wurde nur
manchmal gesehen. Aus der Arbeit von van der Velden kann man
ableiten, das diese wechselnde Wahrnehmung zurückzuführen ist
auf das statistische Verhalten der Licht-quanta, also eine reine
physikalische Angelegenheit ist. Van der Velden schloss aus seinen
Untersuchungen, dass der Sehpurpur der Netzhaut-region 7° nasal
von der Fovea im dunkeladaptierten Auge jeden Beobachters 2
Quanta absorbieren muss und zwar innerhalb 0,02 Sekunden und
eines Sehwinkels von etwa 10′, um eine Lichtempfindung her-
vorzurufen.

Da es somit möglich erscheint eine physikalische Betrachtungs-
weise des Sehvorganges von der physiologischen abzugrenzen, so
kann erwartet werden, dass bei sinngemässer Ausbreitung solcher
Versuche weitere fundamentale Eigentümlichkeiten des Sehorgans
aufgefunden werden können.

Dies war der Zweck der hier beschriebenen Experimente.

Im Kapitel I wird die Zwei-quanten-theorie, die zugehörigen
Grundversuche und unsere experimentelle Bestätigung derselben für
das Stäbchensehen besprochen.

Im Kapitel II werden die Versuche über die Schwellenwerte für
das Stäbchensehen erörtet, in welchen der Einfluss des Sehwinkel
$d$ und der Expositionszeit $t$ untersucht wurde. Es zeigte sich, dass
Abweichungen von der Zweiquantenregel auftreten bei grossen
Werten von $t$ und $d$. Diese Abweichungen können erklärt werden
durch die verminderte Wahrscheinlichkeit des Zustandekommens
einer Lichtempfindung nach dem Beginn des Lichtreizes, welcher
Zustand sich entwickelt nach der Zeit $T$ (ca 0,1 sec) und sich
ausbreitet über die Region $O$ (ca 20′).

Das Kapitel III behandelt die Untersuchungen des Zapfensehens.
Aus Versuchen mit Lichtblitzen an verschiedenen Netzhautstellen
mit mono- und heterochromatischem Licht wird gefolgert, dass jeder
Photorezeptor erregt wird durch die Absorption von e i n e m Quant
und einen Impuls entlang seiner nervösen Verbindung schickt. Eine
Lichtempfindung entsteht, wenn ein zweites Quant absorbiert wird
in einem zweiten Photorezeptor innerhalb des Abstands $D$ vom
ersten und der Zeit $\tau$ nach der Absorption des ersten Quants. Die

Zeit $\tau$ ist in unseren Versuchen unabhängig von der Netzhautstelle und von der Art der errregten Rezeptoren (0,04 sec), sodass möglicherweise die Interaktion zwischen den beiden Impulsen immer geschieht in dem gleichen nervösen Element.

Der Abstand $D$ ist in der Fovea ziemlich unabhängig von der Wellenlänge, sodass $D$ ziemlich gleich ausfällt für verschiedene Zapfensysteme (2-4 Minuten). 7° nasal von der Fovea ist aber für die Zapfen im extremen Rot $D$ 4', für die Stäbchen $D$ 10'.

Für das Zustandekommen einer Lichtempfinding ist es gleichgültig, ob die zwei Quanten in Rezeptoren gleicher Art absorbiert werden oder nicht, sodass eine vollständige gegenseitige Abhängigkeit zwischen Stäbchen und Zäpfen in der Peripherie der Netzhaut, und zwischen den Zapfen der Fovea besteht.

Im Kapitel IV werden die Grundtatsachen der Funktion des nervösen Systems abgehandelt. Es wird erst eine Übersicht der allgemeinen Grundlagen der Histologie des Nervenssystems und der nervösen Übertragung gegeben. In Übereinstimmung mit der durch uns gefundenen gegenseitigen Abhängigkeit zwischen verschiedenen Zapfen der Fovea haben wir angenommen, dass auch wechselseitige Verbindungen zwischen den Zapfen irgendwo in der Sehbahn existieren.

Sherrington's Untersuchungen über die Erregung des einfachen Reflexbogens und Eccles Arbeiten über die Potentialschwankungen in nervösen Elementen beweisen, dass die Forschungsergebnisse der Nervenphysiologie und der physiologischen Optik identisch sind: Das Sehorgan und der einfache Reflexbogen werden erregt durch zwei Impulse.

Im Kapitel V werden Versuche mitgeteilt über die spektrale Empfindlichkeitskurve für Stäbchen und Zapfen. Die Form unserer Foveakurven, erhalten durch Messungen der absoluten Schwellenwerte, unterscheidet sich von der der üblichen Helligkeitskurven. Die weinigen, ziemlich schmallen Gipfel unserer Kurve rühren her von den eigentlichen Empfindlichkeitskurven der einzelnen Rezeptoren. Die Gipfelformen zeigen, dass die Halbwerts-werte dieser Teilkurven ziemlich schmall sind und übereinstimmen mit den Werten der "Modulatoren" von Granit, während sie den weiteren Kurven anderer Autoren als Grundempfindungskurven in Helligkeitmessungen bestimmt, wiedersprechen.

Keine Wellenlängenverschiebung konnte bei Deuteranomalen beobachtet werden, sodass das Sehen anomaler Trichromaten nicht photochemisch gegründet erscheint, sondern nervösen Ursprungs sein muss.

Die Abweichung der "Summationsregel" für kleine Sehwinkel, sowie sie Bouma annahm, muss als verursacht durch die chromatische Aberration des Auges angesehen werden. In der Nähe der absoluten Schwelle können keine fundamentalen Abweichungen vorkommen.

Im Kapitel VI wird das Verhalten der Helligkeitsempfindung studiert mittels des Vergleiches von Lichtblitzen von verschiedener Dauer und Grösse. Für das periphere Sehen muss bei bestimmter Helligkeit die mittlere Zahl der effektiven Absorptionen innerhalb 0,11 Sekunden und eines Areales von 25' Durchmesser für das Stäbchensehen und 4' für das Zapfensehen konstant sein. Es ist wohl sicher, dass die Impulse innerhalb der angegebenen Zeiten und Areale zum Hervorrufen einer Helligkeitsempfindung zusammenarbeiten. Diese Helligkeitsempfindung nimmt in Allgemeinen mit der Zahl der Absorptionen zu. Für das "Rotzapfen" – sehen im Gebiet $7°$ nasal von der Fovea, müssen alle absorbierten Quanten im extremen Rot in einem Zapfen absorbiert werden, da sich dort höchstens ein Rotzapfen befindet, während für das Stäbchensehen dort etwa 400 Stäbchen zur Verfügung sind. Das Gebiet $O$ hat den Charakter einer Rezeptoreinheit, die Zeit $T$ den einer mittleren Zeit.

Das Kapitel VII ist der Sehschärfe geweiht. Insofern der Sehwinkel eines Details der Testfigur gross ist im Vergleich zum Durchmesser der Rezeptoreinheit, insofern verhält sich die Sehschärfe wie die Schwellenwerte und stimmt überein mit dem Zweiquantentheorem. In minimum recognoscibile ist die Sehschärfe proportional der Intensität.

Wenn die Details der Testfigur von der gleichen Grössenordnung sind wie der Durchmesser der Rezeptoreinheit, dann treten Abweichungen von der Zweiquantenregel bei Schwellenwertsbestimmungen auf, weil in einer solchen Einheit nicht verschiedene Lichtempfindungen hervorgerufen werden können.

Im Kapitel VIII werden der simultane und suksessive Vergleich von Lichtempfindungen besprochen und abgeleitet, dass zur Perzeption von Empfindungsunterschieden die erste und letzte Zeit $T$ der Beobachtung und somit für den Sehakt schlechthin die Grenzen des Testfleckes von besonderer Wichtigkeit sind. Die Konsequenzen dieser Tatsachen werden den im Kapitel II erörterten Abweichungen gegenübergestellt als Probleme der Sehschärfe und des Einflusses des Eigengraus der Netzhaut.

Diese „psychologischen" Überlegungen können ein Paraphrase der Impulsübertragung im Nervensystem sein und es wird annehmlich gemacht, dass der „on-Effekt", „off-Effekt" und „on-off-Effekt" der nervösen Übertragung in enger Beziehung steht zu den im letzten Kapitel behandelten Problemen des Sehens.

*Physikalisches Laboratorium Reichs-Universität Utrecht.*
*Die Niederlande.*

# REFERENCES

1. H. A. van de Velden, 1944. Over het aantal lichtquanta, dat nodig is voor een licht prikkel bij het menselijk oog. Physica 11, 179.

   H. A. van de Velden, 1946. The number of quanta necessary for the perception of light of the human eye. Ophthalmologica 111, 321.

2. M. A. Bouman and H. A. van de Velden, 1947. The two quanta-explanation of the dependence of the threshold values and visual acuity on the visual angle and the time of observation. J. Opt. Soc. Am. 37, 908.

3. M. A. Bouman and H. A. van de Velden, 1948. The two quanta hypothesis as a general explanation for the behaviour of threshold values and visual acuity for the several receptors of the human eye, J. Opt. Soc. Am. 38, 570.

4. S. Hecht, 1937. Rods, cones and the chemical basis of vision. Physiol. Rev. 17, 239.

5. T. L. Jahn, 1945. Brightness discrimination and visual acuity as functions of intensity, J. Opt. Soc. Am. 36, 83.

6. R. Bowling Barnes and M. Czerny, 1932. Läszt sich ein Schroteffekt der Photonen mit dem Auge beobachten? Z. Phys. 79, 436.

7. S. L. Polyak, 1941. The retina, Univ. Chicago.

8. G. Østerberg, 1935. Topography of the layer of rods and cones in the human retina. Acta Ophthalm. suppl. 6.

9. H. K. Hartline, 1940. The perceptive field of the optic nerve fibers. Am. J. Phys. 130, 690.

10. G. Wald, 1944. Vision: photochemistry, Medical Physics, Glasser, Chicago.

11. A. F. Bliss, 1948. The mechanism of retinal vitamin A formation, J. Biol. Chem. 172, 165.

12. Hl. de Vries, 1943. The quantum character of light and its bearing upon threshold of vision, the differential sensitivity and visual acuity of the eye, Physica 10, 553.

13. J. M. W. Milatz and L. S. Ornstein, 1935. The electronic excitation-function of the metastable $S_5$ - level of neon. Physica 2, 355.

14. J. M. W. Milatz and N. Bloembergen, 1946. The development of a photo-electric alternating current amplifier with a.c. galvanometer, Physica 11, 449.

15. L. S. Ornstein, W. J. H. Moll, H. C. Burger, 1932. Objektive Spektral-photometrie, Sammlung Vieweg, Braunschweig.

L. S. Ornstein, D. Vermeulen, E. F. M. van der Held, 1930. Calibration of standard lamps for relative and absolute measurements, J. Opt. Soc. Am. 20, 573.

L. S. Ornstein, 1939. Unité de lumière ou méthode standardisée ` de mesure? Revue d'Optique, 12, 385.

16. S. Hecht, S. Shlaer, M. H. Pirenne, 1941. Energy at the threshold of vision. Science 93, 585.

17. E. C. Wassink and M. A. Bouman, 1947. Can phototropism be initiated by a one-quantum-per-cell process? Enzymologica 12, 193.

18. C. Peyrou and H. Piatier, 1946. Emploi de méthodes statistique dans l'étude de la sensibilité de l'oeil. C. F. de Soc. de l'Ac. d. Sc. F 113, 589.

19. G. Wald, 1938. On rhodopsin in solution. J. Gen. Physiol. 21, 795.

20. M. H. Pirenne, 1943. Binocular and uniocular threshold of vision. Nature. 152, 698.

21. J. Steinhardt, 1936. Intensity discrimination in the human eye. J. Gen. Physiol. 20, 185.

22. E. Baumgardt, 1948. Sur la loie spatiale de la brillance liminaire en vision foveale. C. R. de la Soc. de biol. 142, 464.

23. J. F. Schouten and L. S. Ornstein, 1939. Measurements on direct and indirect adaptation by means of a binocular method. J. Opt. Soc. Am. 29, 168.

24. H. Hartridge, 1946. Fixation area in the human eye, Nature 158, 303.

25. W. D. Wright, 1946. Researches on normal and defective colour vision. Kimpton, London.

26. Howell's Textbook of Physiology, 1946. Saunders, London.

27. R. S. Creed, D. Denny-Brown, J. C. Eccles, E. G. T. Lidell, C. S. Sherrington, 1932. Reflex activity of the spinal cord Clarendon press, Oxford.

28. J. C. Eccles, 1946. Synaptic potentials of motoneurons. J. of Neurophysiol. 9, 94.

29. L. C. Thomson and W. D. Wright, 1947. The colour sensitivity of the retina within the central fovea of man. J. Physiol. 105, 316.

30. R. J. Lythgoe, 1937. The absorption spectra of visual purple and of indicator yellow. J. Physiol. 89, 331.

31. E. Ludvigh and E. F. Mc. Carthy, 1938. Absorption of visible light by the refractive media of the human eye. Arch. of Ophthalm. 20, 37.

32. G. Wald, 1945, Human vision and the spectrum, Science 101, 653.

33. L. C. Thomson, 1947. The effect of change of brightness level upon the foveal luminosity curve measured with small fields. J. Physiol 106, 368.

34. Hl. de Vries, 1946. On the basic sensation curves of the three-colour theory. J. Opt. Soc. Am. 36, 121.

    Hl. de Vries, 1947. Kleurenzien. Ned. T. voor Nat. 13, 179.

35. W. S. Stiles, 1939. The directional sensitivity of the retina and the spectral sensitivities of the rods and cones. Proc. Roy. Soc. 127 B, 64.

36. S. Hecht, 1932. Report of a joint discussion on vision. Phys. Soc. London 44, 126.

37. H. V. Walters, 1942. Some experiments on the trichromatic theory of vision. Proc. Roy. Soc. B 131, 27.

38. F. H. G. Pitt, 1944. The nature of normal trichromatic and dichromatic vision. Proc. Roy. Soc. B 132, 101.

39. R. Granit, 1947. Sensory mechanisms of the retina. Oxford University Press.

40. P. J. Bouma, 1942. Die Wahrnehmungsschwelle punktförmiger farbiger Lichter. Physica 9, 890.

41. M. A. Bouman and H. A. van de Velden, 1948. The quanta explanation of vision and the brightness impression for various times of observations and visual angles. J. Opt. Soc. Am. 38, 231.

42. J. ten Doesschate, 1946. Visual acuity and distribution of percipient elements on the retina. Ophthalmologica 112, 1.

43. M. H. Pirenne, 1946. On the variation of visual acuity with light intensity. Proc. Cambr. Phil. Soc. 42, 78.

44. C. H. Graham and C. Cook, 1937. Visual acuity as a function of intensity and exposure time. Am. J. of Psychol. 49, 654.

45. C. H. Graham and R. Margaria, 1935. Area and the intensity-time relation in the peripheral retina. Am. J. Physiol. 113, 299.

46. G. Wald, 1938. Area and visual threshold. J. Gen. Physiol. 21, 269.

47. E. L. Lamar, S. Hecht, S. Shlaer, S. D. Hendly, 1948. Size, shape and contrast of targets by daylight vision. J. Opt. Soc. Am. 38, 741.

48. H. K. Hartline, 1940. The nerve messages in the fibers of the visual pathway. J. Opt. Soc. Am. 30, 239.